Filibert Roth

Timber

An elementary discussion of the characteristics and properties of wood

.

Filibert Roth

Timber
An elementary discussion of the characteristics and properties of wood

ISBN/EAN: 9783337276461

Printed in Europe, USA, Canada, Australia, Japan

Cover: Foto ©Andreas Hilbeck / pixelio.de

More available books at **www.hansebooks.com**

Bulletin No. 10.

U. S. DEPARTMENT OF AGRICULTURE.
DIVISION OF FORESTRY.

TIMBER:

AN ELEMENTARY DISCUSSION OF THE CHARACTERISTICS AND PROPERTIES OF WOOD.

BY

FILIBERT ROTH,
Special Agent in Charge of Timber Physics.

UNDER THE DIRECTION OF
B. E. FERNOW,
CHIEF OF THE DIVISION OF FORESTRY.

WASHINGTON:
GOVERNMENT PRINTING OFFICE.
1895.

TABLE OF CONTENTS.

LIST OF ILLUSTRATIONS.

4

INTRODUCTION.

Wood is now, has ever been, and will continue to be, the most widely useful material of construction. It has been at the base of all material civilization. In spite of all the substitutes for it in the shape of metal, stone, and other materials, the consumption of wood in civilized countries has never decreased; nay, applications in new directions have increased its use beyond the saving effected by the substitutes. Thus, in England, the per capita consumption has increased in the last fifty years more than double, a fact which is especially notable, as the bulk of the timber used there must be imported, while iron and coal are plentiful in Great Britain.

In the United States we can only estimate from the partial data furnished by census returns. By these we find the per capita consumption to have increased for every decade since 1860 at the rate of from 20 to 25 per cent.

Although wood has been in use so long and so universally, there still exists a remarkable lack of knowledge regarding its nature in detail, not only among laymen, but among those who might be expected to know its properties. As a consequence, the practice is often faulty and wasteful in the manner of its use. Experience has been almost the only teacher, and notions—sometimes right, sometimes wrong—rather than well-substantiated facts lead the wood consumer. Iron, steel, and other metals are much better known in regard to their properties than wood. The reason for this imperfect knowledge lies in the fact that wood is not a homogeneous material, like the metals, but a complicated structure, and so variable that one stick will behave very differently from another stick, although cut from the same tree. Not only does the wood of one species differ from that of another, but the butt cut differs from the top log; the heartwood from the sapwood; the wood of the quickly grown sapling of the abandoned field from that of the slowly grown old monarch of the forest. Even the manner in which the tree was sawed and the condition in which the wood was cut and kept influence its behavior and quality. It is, therefore, extremely difficult to study the material for the purpose of establishing general laws, and it becomes necessary to make a specific inspection of the individual stick which is to be applied to a certain purpose. The selection, not only of the most suitable kinds, but of each stick, for the

5

purpose for which it is fit will enter into that improved practice to which we may look both for greater economy and greater efficiency.

The object of this bulletin is to record more systematically than has been done hitherto the knowledge which exists and which will help the wood consumer in the choice of his material and in determining whether, and if so why, a given stick will answer his purpose. Such inspection requires, first, a knowledge of the gross structure and appearance, which give indications of quality and behavior, and then, for finer application, a knowledge of the minute anatomical or microscopic structure. The minute structure will often explain the difference in behavior of various kinds of wood, and a knowledge of it is almost indispensable in distinguishing the various kinds.

In the countries of Europe the kinds of wood used in construction and manufacture are so few that there is but little difficulty in distinguishing them. In our own country the great variety of woods, and of useful woods at that, often makes the mere distinction of the kind or species of tree most difficult. Thus there are at least eight pines (of the thirty-five native ones) in the market, some of which so closely resemble each other in their minute structure that they can hardly be told apart; and yet they differ in quality and should be used separately, although they are often mixed or confounded in the trade. Of the thirty-six oaks, of which probably not less than six or eight are marketed, we can readily recognize by means of their minute anatomy at least two tribes—the white and the black oaks. The distinction of the species is, however, as yet uncertain. The same is true as to the eight kinds of hickory, the six kinds of ash, etc. Before we shall be able to distinguish the wood of these species unfailingly, more study will be necessary. The key given in the present publication, therefore, is by necessity only provisional, requiring further elaboration. It unfortunately had to be based largely on external appearances, which are not always reliable. Sometimes, for general practical purposes, this mere appearance, with some minor attributes, such as color, taste, etc., are together sufficient, especially when the locality is known from which the species came, and in the log pile the determination may by these means be rendered possible when a single detached piece will leave us doubtful as to the species. In the market the distinctions are often most uncertain, and a promiscuous application of names adds to the confusion. To be sure, there is not much virtue in knowing the correct name, except that it assists us in describing the exact kind of material we desire to obtain. Nor is there always much gained in being able to identify the species of wood, but that it predicates certain qualities which are usually found in the species.

In selecting material, then, for special purposes we first determine what species to use as having either one quality which is foremost in our requirements, or several qualities in combination, as shown by actual experience or by experiment.

The uses of the various woods depend on a variety of conditions. The carpenter and builder, using large quantities of material and bestowing a minimum amount of labor on the greater part of the same, uses those kinds which are abundant, and hence cheap, to be had in large dimensions, light to ship, soft to work and to nail, and fairly stiff and insect proof—a combination represented in the conifers. They need not be handsome, hard, tough, or very strong, and may shrink even after they are in place. When it comes to finishing-woods, more stress is laid on color and grain and that the wood shall shrink as little as possible.

The furniture maker, who bestows a maximum amount of work on his material, needs a wood that combines strength, and sometimes toughness, with beauty and hardness, that takes a good polish, keeps joint, and does not easily indent. It must not warp or shrink when once in place, but it need not be light or soft or insect proof or abundant in any one kind, and in large dimensions, nor yet particularly cheap.

Toughness, strength, and hardness combined are sought by the wagon maker. The carriage builder, cooper, and shingle maker look for straight-grained, easy-splitting woods, and for a long fiber, the absence of disturbing resinous and coloring matter, knots, etc. Durability under exposure to the weather, resistance to indentation, and the holding of spikes are required for a good railroad tie; lasting qualities, elasticity, and proportionate dimensions of length and diameter, for telegraph poles.

Sometimes in practice it is immaterial whether the stick be of white oak or red oak, and many wood yards make no distinction, in fact do not know any, but the experienced cooper will quickly distinguish, not by name, perhaps, but by quality, the more porous red or black oak from the less porous white species. On the other hand, the very same white oak—*Quercus alba*, usually a superior article—may furnish so poor material for a handle or a plow beam that a stick of red oak would be preferable. The inspection, then, must be made not only for the species but for the quality, with reference to the purpose for which the stick is to be used.

That the inspection should have regard to defects and unhealthy condition (often indicated by color) goes without saying, and such inspection is usually practiced. That knots, even the smallest, are defects which for some uses condemn the material altogether needs hardly to be mentioned, but that season checks, even those that have closed by subsequent shrinkage, remain elements of weakness is not so readily appreciated. Yet there can not be any doubt of this, since the intimate connection of the wood fibers, once interrupted, is never reestablished. The careful wood user, therefore, is concerned as to the manner in which his material was treated after the felling, for according to the more or less careful seasoning of it the season checks, not altogether avoidable, are more or less abundant. This is practically recognized by splitting wagon and cooperage stock in the woods and seasoning it partly shaped, and also in making a distinction, often unnecessarily, between air-dried and kiln-dried material.

Where strength is required, the weight of the material will give good indications, for it is now pretty well established that weight and strength go more or less together. But since weight in the green wood is made up of at least three elements, namely, that of the wood fiber itself, that of the water in the cell spaces, and that of the water in the cell walls, the weight is deceptive unless we know also the moisture condition of the stick or else ascertain the specific weight of the dry wood. That the moisture contents influence considerably the strength of the material is now well proven, strength increasing with loss of moisture, and hence in practice allowance should be made according to whether the stick is to be used where it will be exposed to the weather or under cover and painted.

In some woods like the pines and the "ring porous" woods, such as oak, chestnut, and hickory, in which each annual layer or ring is made up of two distinct parts, the loose, porous spring wood and the dense and firm summer wood, the proportion of the latter per square inch of cross section—usually but not always depending on the width of the ring—furnishes a more direct criterion than the weight alone. The color effect of itself gives indications of the weight, since both weight and color effect depend on the same feature, namely, quantity of material; hence the larger quantity of dense summer wood on the cross section occasions darker color, which is usually indicative of strength. Color, too, must be consulted to detect incipient decay. Again, the difference in firmness and hardness of the summer wood itself, as tested by the knife or recognized in the difference of color effect by the practiced eye, furnishes another criterion in the selection of the stick.

Lastly, the manner in which the stick is sawed from the tree has a remarkable influence upon its qualities and behavior, and it should, therefore, either be specially sawed or selected with a view to its character and to the purpose for which it is to be used. This is a matter fully appreciated among only a few wood users, like the wheelwrights, piano makers, etc., but it needs to be observed much more than it is, even in building. Quarter or rift sawing, i. e., cutting sticks or boards out of the log in such a manner that the annual rings are cut through as nearly as possible radially, has lately been practiced largely for the sake of the beauty of the even grain thus obtained, and also for flooring on account of the better wear which the even exposure of the grain (hard bands of summer wood on edge) secures; but it should be much more widely applied to secure greater strength and more uniform seasoning and thus to reduce to some extent the one drawback to wood as a material of construction, that is, its liability to "working" (shrinking and swelling). The reason for the superiority of quarter-sawed pieces, as well as the general fact that the manner of sawing out a stick affects the general character and behavior of the same, will appear from the following considerations :

A square column or beam cut so as to contain the heart or pith of the tree in its center—which, by the way, is the weakest part on

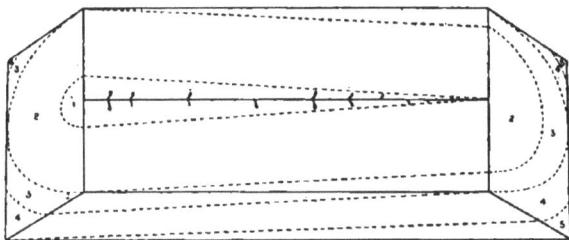

account of the many knots which it invariably and necessarily con-tains—consists in the main of five structural aggregates (see fig. 1), namely: (1) In the center a cone of wood fibers with the base in the butt end and the apex in the top end, the base representing the rings of as many years as it took the tree to attain the height of the column; none of the fibers belonging to these rings appear in the top section excepting those of the last ring which forms the apex of the cone; (2) a hollow cylinder of material surrounding the cone, all fibers of which are found in both sections and con-tinuously through the whole length of the column; all the entire rings at the bottom belong in this cylinder, and undoubtedly form the strongest part of the column; (3) surrounding this cylinder a partial cylindrical envelope of wood fibers, all of which are represented in the top section, but only a part appear at the corners of the bottom; most of them, there fore, do not run through the whole length, but are cut through at vary-ing lengths, thereby presenting the "bastard faces" on the sides of the column; (4) a partial envelope whose

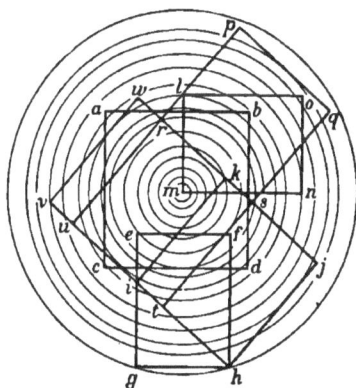

Fig. 2.—Possibilities of cutting timber from a log with reference to position of grain.

radial extent is limited by the corners of the basal section, imperfect at both ends; (5) the corners at the top, three-sided pyramids with the base in the top section, the fibers running out at varying lengths.

Now, it will be readily admitted that each of these "structural aggre-gates" has a different value in the combined strength of the whole. If the stick be cut with the center or pith in one side (see fig. 2) all these aggregates will be halved; if the stick be cut out differently, for instance, with the heart entirely out or if it be made longer or shorter, or rectangular instead of square, in each case the proportion of each of the aggregates changes, and hence it stands to reason that the strength of the column, or beam, or stick, changes according to the manner in which it is cut from the tree. This most evident and important fact has, it seems, escaped our best engineers and experimenters, who have tested beams without taking account of this disturbing element, and it is certainly overlooked most generally by builders and carpenters in their selection of material.

While it may perhaps not be expected that the sawing at the mill will be done with more care so as to secure the best results in application, or that the special advantage of quarter sawing will soon be sufficiently appreciated so as to extend its use in such a manner that the greater efficiency of the quarter sawed material will compensate for the greater expense of the operation, wood users may at least be expected to make their selections from the sawed material in the yard, and shape it for their particular use with greater care.

There is no country in which wood is more lavishly used than in the United States, and none in which nature has more bountifully provided for all reasonable requirements. In the absence of proper efforts to secure reproduction, the most valuable kinds are rapidly being decimated, and the necessity of a more rational and careful use of what remains is clearly apparent. By greater care in selection, however, not only can the duration of the supply be extended, but more satisfactory results will accrue from its use.

B. E. FERNOW.

WASHINGTON, D. C., *September 15, 1895.*

TIMBER.

CHARACTERISTICS AND PROPERTIES OF WOOD.

I.—STRUCTURE AND APPEARANCE.

The structure of wood affords the only reliable means of distinguish ing the different kinds. Color, weight, smell, and other appearances, which are often direct or indirect results of structure, may be helpful in this distinction but can not be relied upon entirely. In addition, structure underlies nearly all the technical properties of this important product and furnishes an explanation why one piece differs as to these properties from another.

Structure explains why oak is heavier, stronger, and tougher than pine; why it is harder to saw and plane, and why it is so much more difficult to season without injury. From its less porous structure alone, it is evident that a piece of a young and thrifty oak is stronger than the porous wood of an old or stunted tree; or that Georgia or longleaf pine excels white pine in weight and strength. Keeping especially in mind the arrangement and direction of the fibers of wood, it is clear at once why knots and "crossgrains" interfere with the strength of timber.

It is due to structural peculiarities that "honeycombing" occurs in rapid seasoning, that "checks" or cracks extend radially and follow pith rays, that tangent or "bastard" boards shrink and warp more than quartered lumber. These same peculiarities enable cherry and oak to take a better finish than basswood or coarse grained pine.

Moreover, structure, aided by color, determines the beauty of wood. All the pleasing figures, whether in a hard-pine ceiling, a desk of quartered oak, or in the beautiful panels of "curly" or "bird's-eye" maple decorating the saloon of a ship or a palace car, are due to differences in the structure of the wood. Knowing this, the appearance of any particular section can be foretold, and almost unlimited choice and combination are thereby suggested.

Thus a knowledge of structure not only enables us to distinguish the different woods, judge as to their qualities, and explain the causes of their beauty, but it also becomes an invaluable aid to the thoughtful worker, guiding him to a more careful selection and a more perfect use of his material.

CLASSES OF TREES.

The timber of the United States is furnished by three well-defined classes of trees: the needle-leaved, naked-seeded conifers (pine, cedar, etc.), the dicotyledonous (with two seed leaves), broad-leaved trees (oak, poplar, etc.), and to an inferior extent by the monocotyledonous (with one seed leaf), palms, yuccas, and their allies, which last are confined to the most southern parts of the country.

Broad-leaved trees are also known as deciduous trees, although especially in warm countries, many of them are evergreen,[1] while the conifers are commonly termed "evergreens," although the larch, bald cypress, and others shed their leaves every fall, and even the names "broad-leaved" and "coniferous," though perhaps the most satisfactory, are not at all exact, for the conifer ginkgo has broad leaves and bears no cones.

In the lumber trade, the woods of broad-leaved trees are known as "hardwoods," though poplar is as soft as pine, and the coniferous woods are "soft woods," notwithstanding that yew ranks high in hardness even when compared to "hardwoods."

Both in the number of different kinds of trees or species and still more in the importance of their product the conifers and broad-leaved trees far excel the palms and their relatives.

In the manner of growth both conifers and broad-leaved trees behave alike, adding each year a new layer of wood which covers the old wood in all parts of the stem and limbs. Thus the trunk continues to grow in thickness throughout the life of the tree by additions (annual rings) which in temperate climates are, barring accidents, accurate records of the tree. With the palms and their relatives the stem remains generally of the same diameter, the tree of a hundred years being as thick as it was at ten years, the growth of these being only at the top. Even where a peripheral increase takes place, as in the yuccas, the wood is not laid on in well-defined layers; the structure remains irregular throughout.

Though alike in their manner of growth, and therefore similar in their general make-up, conifers and broad-leaved trees differ markedly in the details of their structure and the character of their wood. The wood of all conifers is very simple in its structure, the fibers composing the main part of the wood being all alike and their arrangement regular. The wood of broad-leaved trees is complex in structure; it is made up of several different kinds of cells and fibers and lacks the regularity of arrangement so noticeable in the conifers. This difference is so great that in a study of wood structure it is best to consider the two kinds separately.

WOOD OF CONIFEROUS TREES.

Examining a smooth cross section or end face of a well-grown log of Georgia pine or Norway pine, we distinguish an envelope of reddish, scaly bark, a small whitish pith at the center, and between these the wood in a great number of concentric rings.

---

[1] In Ceylon even the cultivated cherry has become an evergreen.

The bark of a pine stem is thickest and roughest near the base, decreases rapidly in thickness from 1½ inches at the stump to one-tenth inch near the top of the tree, and forms in general about 10 to 15 per cent of the entire trunk.

The pith is quite thick, usually one-eighth to one-fifth inch in Norway pine and in the southern species, though much less so in white pine, and is very thin, one-fifteenth to one twenty-fifth inch in cypress, cedar, and larch.

In woods with a thick pith, this latter is finest at the stump, grows rapidly thicker upward, and becomes thinner again in the crown and limbs, the first 1 to 5 rings adjoining it behaving similarly.

A zone of wood next to the bark, 1 to 3 or more inches wide, and containing 30 to 50 or more annual rings, is of lighter color; this is the sapwood, the inner, darker part of the log being the heartwood. In the former many cells are active and store up starch and otherwise assist in the life processes of the tree, although only the last or outer layer of cells the cambium, forms the growing part and the true life of the tree. In the heartwood all cells are lifeless cases, and serve only the mechanical function of keeping the tree from breaking under its own great weight, or from being laid low by the winds.

The darker color of the heartwood is due to infiltration of chemical substances into the cell walls, but the cavities of the cells in pine are not filled up, as is sometimes believed, nor do their walls grow thicker, nor is their wall any more lignified than in the sapwood. Sapwood varies in width and in the number of rings which it contains, even in different parts of the same tree; the same year's growth which is sapwood in one part of a disk may be heartwood in another. Sapwood is widest in the main part of the stem and varies often within considerable limits, and without apparent regularity. Generally it becomes narrower toward the top and in the limbs, its width varying with the diameter, and being least, in a given disk, on the side which has the shortest radius. Sapwood of old and stunted pines is composed of more rings than that of young and thrifty specimens. Thus in a pine 250 years old, a layer of wood or annual ring does not change from sapwood to heartwood until seventy or eighty years after it is formed, while in a tree 100 years old, or less, it remains sapwood only from thirty to sixty years. The width of the sapwood varies considerably for different kinds of pines; it is small for longleaf and white pine, and great for loblolly and Norway pines. Occupying the peripheral part of the trunk the proportion which it forms of the entire mass of the stem is always great. Thus even in old trees of longleaf pine the sapwood forms about 40 per cent of the merchantable log, while in the loblolly and in all young trees the bulk of the wood is sapwood.

The concentric, annual, or yearly rings, which appear on the end face of a log are cross sections of so many thin layers of wood. Each such layer forms an envelope around its inner neighbor, and is in turn covered by the adjoining layer without, so that the whole stem is built up of a series of thin hollow cylinders, or rather cones. A new layer of wood is formed each season, covering the entire stem, as well as all the living branches. The thickness of this layer, or the width of the yearly ring, varies greatly in different trees and also in different parts of the same tree. In a normally grown, thrifty pine log the rings are widest near the pith, growing more and more narrow toward the bark. Thus the central 20 rings in a disk of an old longleaf pine may each be one-eighth to one-sixth inch (3 to 4 mm.) wide, while the 20 rings next to the bark may average only one-thirtieth inch (0.7 mm.). In our forest trees rings of one-half inch in width occur only near the center in disks of very thrifty trees of both conifers and hard woods; one-twelfth inch represents good thrifty growth, and the minimum width of about one two-hundredths inch (0.2 mm.) is often seen in stunted spruce and pine. The average width of rings in well-grown old white pine will vary from one-twelfth to one-eighteenth inch, while in the slower growing longleaf pine it may be one twenty-fifth to one-thirtieth of an inch. The same layer of wood is widest near the stump in very thrifty young trees, especially if grown in the open park, but in old forest trees the same year's growth is wider in the upper part of the tree, being narrowest near the stump and often also near the very tip of the stem. Generally the rings are widest near the center, growing narrower towards bark. In logs from stunted trees the order is often reversed, the interior rings being thin and the outer rings widest. Frequently, too, zones or bands of very narrow rings, representing unfavorable periods of growth, disturb the general regularity. Few trees, even among pines, furnish a log with truly circular cross section; usually it is an oval, and at the stump commonly quite an irregular figure. Moreover, even in very regular or circular disks the pith is rarely in the center, and frequently one radius is conspicuously longer than its opposite, the width of some of the rings, if not all, being greater on one side than on the other. This is nearly always so in the limbs, the lower radius exceeding the upper.

In extreme cases, especially in the limbs, a ring is frequently conspicuous on one side and almost or entirely lost to view on the other. Where the rings are extremely narrow, the dark portion of ring is often wanting, the color being quite uniform and light. The greater regularity or irregularity of the annual rings has much to do with the technical qualities of the timber.

SPRING AND SUMMER WOOD.

Examining the rings more closely, it is noticed that each ring is made up of an inner, softer, light-colored, and an outer, or peripheral, firmer and darker-colored portion. Being formed in the fore part of the season, the inner, light-colored part is termed spring wood, the outer, darker portion being the summer wood of the ring. Since the latter is very heavy and firm, it determines to a large extent the weight and strength of the wood, and as its darker color influences the shade of color of the entire piece of wood, this color effect becomes a valuable aid in distinguishing heavy and strong from light and soft pine wood. In most hard pines, like the longleaf, the dark summer wood appears as a distinct band, so that the yearly ring is composed of two sharply defined bands—an inner, the spring wood, and an outer, the summer wood. But in some cases, even in hard pines, and normally in the wood of white pines, the spring wood passes gradually into the darker summer wood, so that a sharply defined line occurs only where the spring wood of one ring abuts against the summer wood of its neighbor. It is this clearly defined line which enables the eye to distinguish even the very narrow rings in old pines and spruces. In some cases, especially in the trunks of Southern pines, and normally on the lower side of pine limbs, there occur

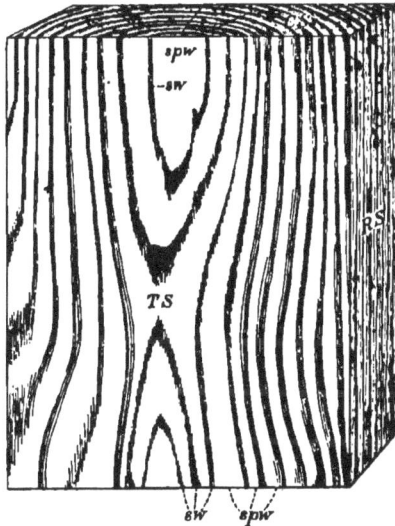

FIG. 3.—Board of pine. *CS*, cross section; *RS*, radial section: *TS*, tangential section; *sw*, summer wood; *spw*, spring wood.

dark bands of wood in the spring wood portion of the ring, giving rise to false rings which mislead in a superficial counting of rings. In the disks cut from limbs these dark bands often occupy the greater part of the ring and appear as "lunes" or sickle-shaped figures. The wood of these dark bands is similar to that of the true summer wood—the cells have thick walls, but usually lack the compressed or flattened form.

Normally, the summer wood forms a greater proportion of the ring in the part of the tree formed during the period of thriftiest growth. In an old tree this proportion is very small in the first 2 to 5 rings about the pith, and also in the part next to the bark, the intermediate part showing a greater proportion of summer wood. It is also greatest in a disk taken from near the stump and decreases upward in the stem, thus fully accounting for the difference in weight and firmness of the wood of these different parts. In the longleaf pine the summer wood often forms scarcely 10 per cent of the wood in the central 5 rings; 40 to 50 per cent of the next 100 rings; about 30 per cent in the next 50, and only about 20 per cent in the 50 rings next to the bark. It averages 45 per cent of the wood of the stump and only 24 per cent of that of the top.

Sawing the log into boards, the yearly rings are represented on the board faces of the middle board (radial sections) by narrow, parallel stripes (see fig. 3), an inner, lighter stripe, and its outer, darker neighbor always corresponding to one annual ring.

On the faces of the boards nearest the slab (tangential or "bastard" boards) the several years' growth should also appear as parallel, but much broader stripes. This they do only if the log is short and very perfect. Usually a variety of pleasing patterns is displayed on the boards, depending on the position of the saw cut, and on the regularity of growth of the log. (See fig. 3.)

Where the cut passes through a prominence (bump or crook) of the log, irregular, concentric circlets and ovals are produced, and on almost all tangent boards, arrow, or V-shaped forms occur.

ANATOMICAL STRUCTURE.

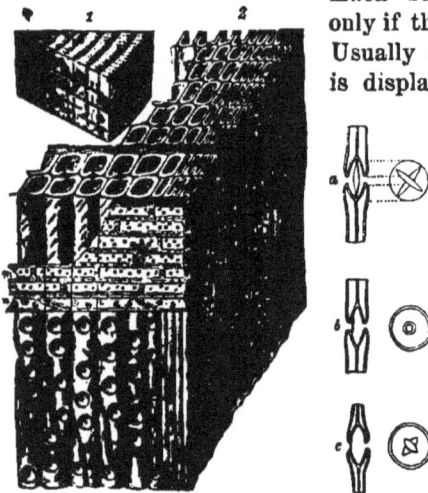

Fig. 4.—Wood of spruce. 1, natural size; 2, small part of one ring magnified 100 times. The vertical tubes are wood fibers, in this case all "tracheids." *m*, medullary or pith ray; *n*, transverse tracheids of pith ray: *a, b,* and *c,* bordered pits of the tracheids, more enlarged.

Holding a well-smoothed disk, or cross section one-eighth inch thick toward the light, it is readily seen that pine wood is a very porous structure. If viewed with a strong magnifier, the little tubes, especially in the spring wood of the rings, are easily distinguished and their arrangement in regular straight radial rows is apparent. Scattered through the summer wood portion of the rings, numerous irregular grayish dots (the resin ducts) disturb the uniformity and regularity of the structure. Magnified 100 times, a piece of spruce, which is similar to pine, presents a picture like that shown in fig. 4. Only short pieces of the tubes or cells of which the wood is composed are represented in the picture.

The total length of these fibers is one-twentieth to one-fifth inch, being smallest near the pith, and is 50 to 100 times as great as their width (fig. 5). They are tapered and closed at their ends, polygonal, or rounded and thin walled, with large cavity, lumen or internal space in the spring wood, thick walled and flattened radially with the internal space or lumen much reduced in the summer wood. (See right-hand portion of fig. 4). This flattening, together with the thicker walls of the cells which reduces the lumen, causes the greater firmness and darker color of the summer wood—there is more material in the same volume. As shown in the figure, the tubes, cells, or "tracheids" are decorated on their walls by circlet-like structures, the "bordered pits," sections of which are seen more magnified at *a*, *b*, and *c*, fig. 4. These pits are in the nature of pores, covered by very thin membranes, and serve as waterways between the cells or tracheids.

The dark lines on the side of the smaller piece (1, fig. 4) appear when magnified (in 2, fig. 4) as tiers of 8 to 10 rows of cells, which run radially (parallel to the rows of tubes or tracheids) and are seen as bands on the radial face and as rows of pores on the tangential face. These bands or tiers of cell rows are the medullary rays or pith rays, and are common to all our lumber woods. In the pines and other conifers they are quite small, but they can readily be seen, even without a magnifier, if a radial surface of split wood (not smoothed) is examined. The entire radial face will be seen almost covered with these tiny structures, which appear as fine but conspicuous cross lines. As shown in fig. 4 the cells of the medullary or pith rays are smaller and very much shorter than the wood fibers or tracheids and their long axis is at right angles to that of the fibers. In pines and spruces the cells of the upper and lower rows of each tier or pith ray have "bordered" pits like those of the wood fibers or tracheids proper, but the cells of the intermediate rows, and of all rows in the rays of cedars, etc., have only "simple" pits, i. e., pits levoid of the saucer-like "border" or rim.

In pine, many of the pith rays are larger than the majority, each containing a whitish line, the horizontal resin duct, which, though much smaller, resembles the vertical ducts seen on the cross section. The larger vertical resin ducts are best observed on

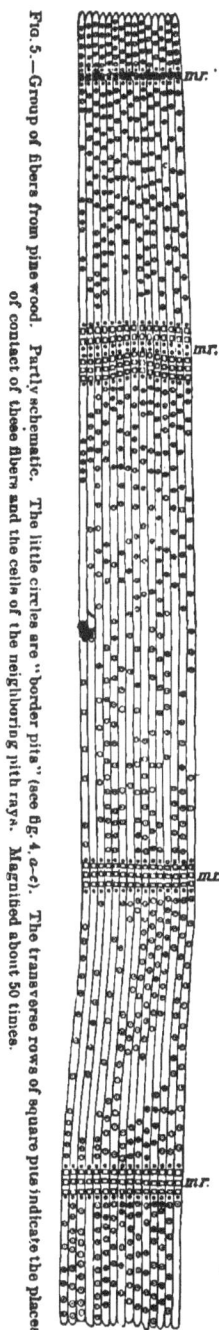

Fig. 5.—Group of fibers from pine wood. Partly schematic. The little circles are "border pits" (see fig. 4, *a–c*). The transverse rows of square pits indicate the places of contact of these fibers and the cells of the neighboring pith rays. Magnified about 50 times.

removal of the bark from a fresh piece of white pine, cut in winter, where they appear as conspicuous white lines, extending often for many inches up and down the stem.

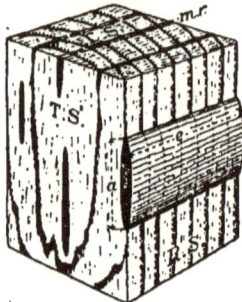

Neither the horizontal nor the vertical resin ducts are vessels or cells, but are openings between cells, i. e., intercellular spaces, in which the resin accumulates, freely oozing out when the ducts of a fresh piece of sapwood are cut. They are present only in our coniferous woods, and even here they are restricted to pine, spruce, and larch, and are normally absent in fir, cedar, cypress, and yew.

Altogether the structure of coniferous wood is very simple and regular, the bulk being made up of the small fibers called tracheids, the disturbing elements of pith rays and resin ducts being insignificant, and hence the great uniformity and great technical value of coniferous wood.

Fig. 6.—Block of oak. *C. S.*, cross section; *R. S.*, radial section; *T. S.*, tangential section; *m. r.*, medullary or pith ray; *a*, height, *b*, width, and *e*, length of a pith ray.

WOOD OF BROAD-LEAVED TREES.

On a cross section of oak, the same arrangement of pith and bark, of sapwood and heartwood, and the same disposition of the wood in well-defined concentric or annual rings occurs, but the rings are marked by lines, or rows, of conspicuous pores or openings which occupy the greater part of the spring wood of each ring (see fig. 6, also fig. 8) and are, in fact, the hollows of vessels through which the cut has been made. On the radial section, or quarter-sawed board, the several layers appear as so many parallel stripes (see fig. 7); on the tangential section or "bastard" face, patterns similar to those mentioned for pine wood are observed. But while the patterns in hard pine are marked by the darker summer wood and are composed of plain, alternating stripes of darker and lighter wood, the figures in oak (and other broad-leaved woods) are due chiefly to the vessels,

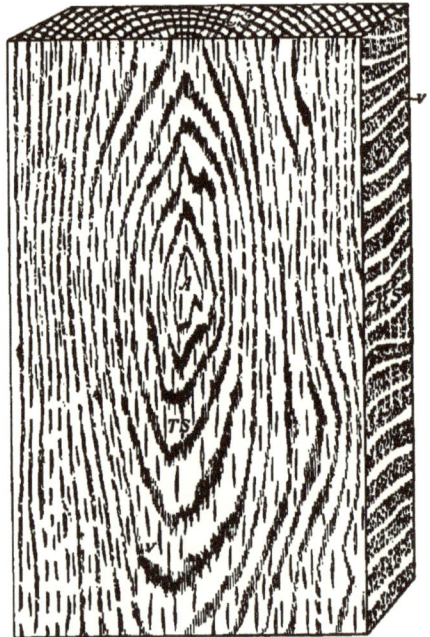

Fig. 7.—Board of oak. *CS*, cross section; *RS*, radial section; *TS*, tangential section; *v*, vessels or pores, cut through; *A*, slight curve in log which appears in section as an islet.

those of the spring wood in oak being the most conspicuous (see fig. 7); so that in an oak table the darker, shaded parts are the spring wood, the lighter, unicolored parts the summer wood.

On closer examination of the smoothed cross section of oak, the spring wood part of the ring is found to be formed, in great part, of pores: large, round, or oval openings made by the cut through long

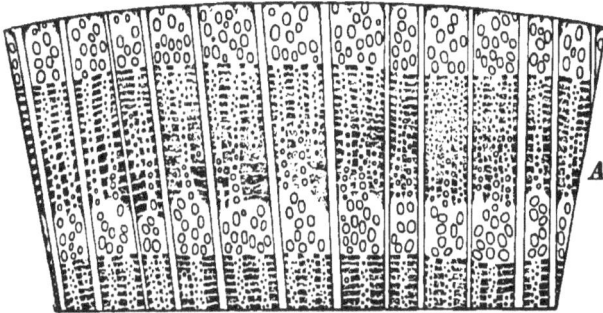

Fig. 8 A.—Cross section of oak magnified about 5 times.

vessels. These are separated by a grayish and quite porous tissue (see fig. 8 A), which continues here and there in the form of radial, often branched, patches (not the pith rays) into and through the summer wood to the spring wood of the next ring. The large vessels of the spring wood, occupying 6 to 10 per cent of the volume of a log in very good oak, and 25 per cent or more in inferior and narrow-ringed lumber, are a very important feature, since it is evident that the greater their share in the volume, the lighter and weaker the wood. They are smallest near the pith, and grow wider outward; they are wider in the stem than limb and seem to be of indefinite length, forming open channels in some cases probably as long as the tree itself.

Scattered through the radiating gray patches of porous wood are vessels similar to those of the spring wood, but decidedly smaller. These vessels are usually fewer and larger near the spring wood, and smaller and more numerous

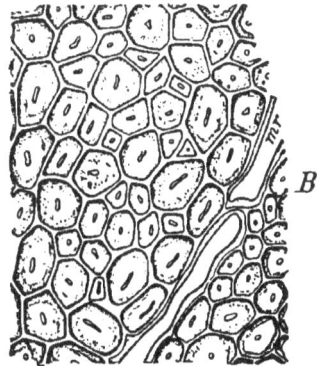

Fig. 8 B.—Portion of the firm bodies of fibers with two cells of a small pith ray *mr*. Highly magnified.

in the outer portions of the ring. Their number and size can be utilized to distinguish the oaks classed as white oaks from those classed as black and red oaks; they are fewer and larger in red oaks, smaller but much more numerous in white oaks. The summer wood, except for these radial grayish patches, is dark colored and firm. This firm portion, divided into bodies or strands by these patches of porous wood and also by fine wavy concentric lines of short, thin-walled cells (see fig. 8 A), consists of thick-walled fibers (see fig. 8 B) and is the chief element of strength in oak wood. In good white oak it forms one-half and more of the wood; it cuts like horn, and the cut surface is shiny and of a deep chocolate-brown color. In very narrow-ringed wood and in inferior red oak it is usually much reduced in quantity as well as quality.

The pith rays of the oak, unlike those of coniferous woods, are at least in part very large and conspicuous (see fig. 6, their height indicated by the letter *a*, and their width by the letter *b*). The large medullary rays of oak are often twenty and more cells wide and several hundred cell rows in height, which amount commonly to one or more inches. These large rays are conspicuous on all sections. They appear as long, sharp, grayish lines on the cross section, as short, thick lines, tapering at each end, on the tangential or "bastard" face, and as broad, shiny bands, the "mirrors," on the radial section. In addition to these coarse rays, there is also a large number of small pith rays, which can be seen only when magnified. On the whole, the pith rays form a much larger part of the wood than might be supposed. In specimens of good white oak it has been found that they formed about 16 to 25 per cent of the wood.

MINUTE STRUCTURE.

If a well-smoothed, thin disk, or cross section of oak (say one-sixteenth inch thick) is held up to the light, it looks very much like a sieve, the pores or vessels appearing as clean-cut holes; the spring wood and gray patches are seen to be quite porous, but the firm bodies of fibers between them are dense and opaque. Examined with the magnifier it

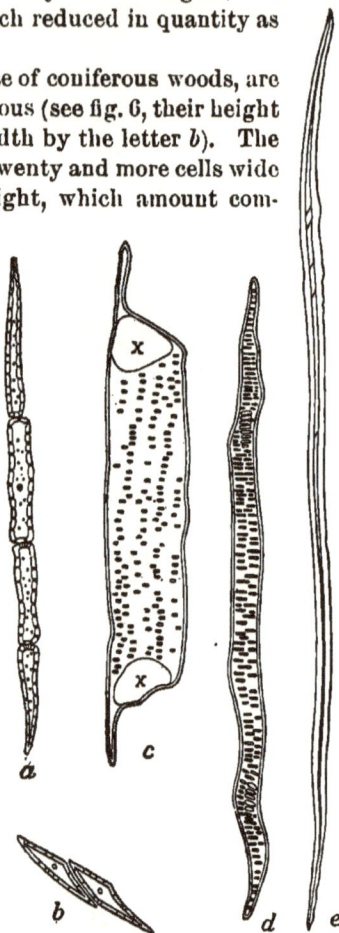

FIG. 9.—Isolated fibers and cells. *a*, four cells of wood parenchyma; *b*, two cells from a pith ray; *c*, a single joint or cell of a vessel, the openings *x* leading into its upper and lower neighbors; *d*, tracheid; *e*, wood fiber proper.

will be noticed that there is no such regularity of arrangement in straight rows as is conspicuous in the pine; on the contrary, great irregularity prevails. At the same time, while the pores are as large as pin holes, the cells of the denser wood, unlike those of pine wood, are too small to be distinguished. Studied with the microscope, each vessel is found to be a vertical row of a great number of short, wide tubes, joined end to end (fig. 9, c). The porous spring wood and radial gray tracts are partly composed of smaller vessels, but chiefly of tracheids like those of pine, and of shorter cells, the "wood paren-chyma," resembling the cells of the medullary rays. These latter, as well as the fine concentric lines mentioned as occurring in the summer wood, are composed entirely of short, tube-like parenchyma cells with square or oblique ends (fig. 9, a and b). The wood fibers proper, which form the dark, firm bodies referred to, are very fine, thread-like cells one twenty-fifth to one-tenth inch long, with a wall commonly so thick that scarcely any empty internal space or lumen remains (figs. 9, e, and 8, B).

If instead of oak a piece of poplar or basswood (fig. 10) had been used in this study, the structure would have been found to be quite different. The same kinds of cell-elements, vessels, etc., are, to be

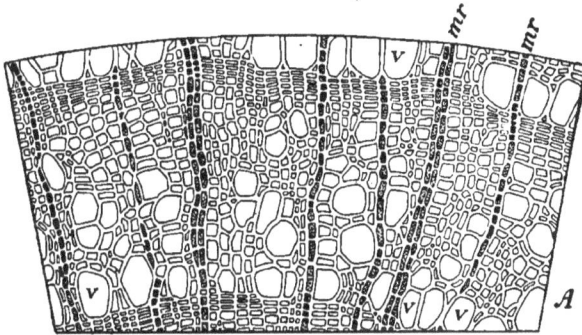

FIG. 10.—Cross section of basswood (magnified). v, vessels; mr, pith rays.

sure, present, but their combination and arrangement is different, and thus from the great variety of possible combinations results the great variety of structure and, in consequence, of the qualities which distin-guish the wood of broad-leaved trees. The sharp distinction of sap-wood and heartwood is wanting; the rings are not so clearly defined, the vessels of the wood are small, very numerous, and rather evenly scattered through the wood of the annual ring, so that the distinction of the ring almost vanishes and the medullary or pith rays, in poplar, can be seen, without being magnified, only on the radial section.

DIFFERENT GRAIN OF WOOD.

The terms "fine grained," "coarse grained," "straight grained" and "cross grained" are frequently applied in woodworking. In com-mon usage, wood is "coarse grained" if its annual rings are wide, "fine grained" if they are narrow; in the finer wood industries a "fine-grained" wood is capable of high polish while a "coarse-grained" wood

is not, so that in this latter case the distinction depends chiefly on hardness, and in the former on an accidental case of slow or rapid growth.

Generally the direction of the wood fibers is parallel to the axis of
the stem or limb in which they occur, the wood is straight grained, but

FIG. 11.—Spiral grain.
Season checks, after re-
moval of bark, indicate
the direction of the fibers
or grain.

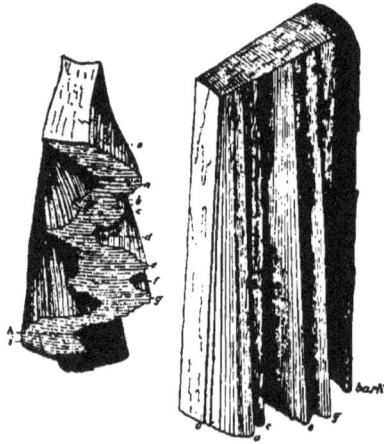

FIG. 12.—Alternating spiral grain in cypress. Side
and end view of same piece. When the bark was
at *o* the grain at this point was straight. From
that time each year it grew more oblique in one
direction, reaching a climax at *a*, and then turned
back in the opposite direction. These alternations were repeated periodically, the bark sharing
in these changes.

in many cases the course of the fibers is spiral or twisted around the
tree as shown in fig. 11, and sometimes (commonly in butts of gum
and cypress) the fibers of several layers are oblique in one direction,
and those of the next series of layers are oblique in the opposite
direction, as shown in fig. 12; the wood is
cross or twisted grained. Wavy grain in a
tangential plain as seen on the radial section
is illustrated in fig. 13, which represents an
extreme case observed in beech. This same
form also occurs on the radial plain, causing
the tangential section to appear wavy or in
transverse folds. When wavy grain is fine,
i. e., the folds or ridges small but numerous,
it gives rise to the "curly" structure frequently seen in maple. Ordinarily, neither
wavy, spiral, nor alternate grain is visible
on the cross section; its existence often escapes the eye even on smooth,
longitudinal faces in sawed material, so that the only safe guide to
their discovery lies in splitting the wood in the two normal plains.

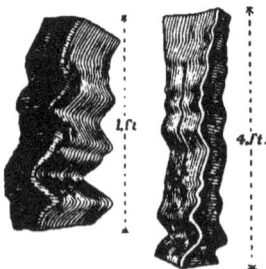

FIG. 13.—Wavy grain in beech;
after Nördlinger.

Generally the surface of the wood under the bark, and therefore also
that of any layer in the interior, is not uniform and smooth, but is

channeled and pitted by numerous depressions which differ greatly in size and form. Usually, any one depression or elevation is restricted to one or few annual layers (i. e., seen only in one or few rings) and is then lost, being compensated (the surface at the particular spot evened up) by growth. In some woods, however, any depression or elevation once attained grows from year to year and reaches a maximum size which is maintained for many years, sometimes throughout life.

In maple, where this tendency to preserve any particular contour is very great, the depressions and elevations are usually small (commonly less than one-eighth inch), but very numerous. On tangent boards of such wood the sections of these pits and prominences appear as circlets and give rise to the beautiful "bird's-eye" or "landscape" structure. Similar structures in the burls of black ash, maple, etc., are frequently due to the presence of dormant buds, which cause the surface of all the layers through which they pass to be covered by small conical elevations, whose cross sections on the sawed board appear as irregular circlets or islets each with a dark speck, the section of the pith or "trace" of the dormant bud in the center.

In the wood of many broad-leaved trees the wood fibers are much longer when full grown than when they are first formed in the cambium or growing zone. This causes the tips of each fiber to crowd in between the fibers above and below, and leads to an irregular interlacement of these fibers, which adds to the toughness but reduces the cleavability of the wood.

At the junction of limb and stem the fibers on the upper and lower sides of the limb behave differently. On the lower side they run from the stem into the limb, forming an uninterrupted strand or tissue and a perfect

Fig. 14.—Section of wood showing position of the grain at base of a limb. *P*, pith of both stem and limb; 1-7, seven yearly layers of wood; *a, b*, knot or basal part of a limb which lived four years, then died and broke off near the stem, leaving the part to the left of *a, b*, a "sound" knot, the part to the right a "dead" knot, which would soon be entirely covered by the growing stem.

union. On the upper side the fibers bend aside, are not continuous into the limb, and hence the connection is imperfect (fig. 14).

Owing to this arrangement of the fibers, the cleft made in splitting never runs into the knot, if started on the side above the limb, but is apt to enter the knot if started below, a fact well understood in wood craft. When limbs die, decay, and break off, the remaining stubs are surrounded and may finally be covered by the growth of the trunk, and thus give rise to the annoying "dead" or "loose" knots.

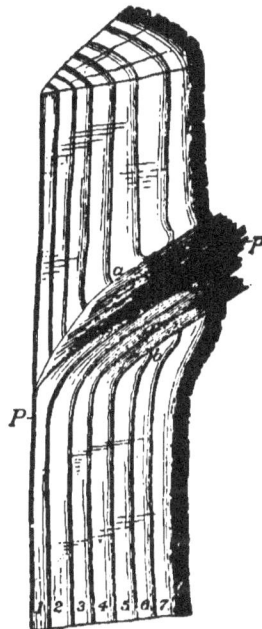

COLOR AND ODOR.

Color, like structure, lends beauty to the wood, aids in its identifica-tion, and is of great value in the determination of its quality. Con-sidering only the heartwood, the black color of the persimmon, the dark brown of the walnut, the light brown of the white oaks, the red-dish brown of the red oaks, the yellowish white of the tulip and poplar, the brownish red of the redwood and cedar, the yellow of the papaw and sumac, are all reliable marks of distinction; and color together with luster and weight are only too often the only features depended upon in practice. Newly formed wood, like that of the outer few rings, has but little color. The sapwood generally is light, and the wood of trees which form no heartwood changes but little, except when stained by forerunners of disease.

The different tints of colors, whether the brown of oak, the orange brown of pine, the blackish tint of walnut, or the reddish cast of cedar, are due to pigments, while the deeper shade of the summer-wood bands in pine and cedar, or in oak or walnut, is due to the fact that the wood being denser, more of the colored wood substance occurs on a given space, i. e., there is more colored matter per square inch.

Wood is translucent, a thin disk of pine permitting light to pass through quite freely. This translucency affects the luster and bright-ness of lumber. When wood is attacked by fungi it becomes more opaque, loses its brightness, and in practice is designated "dead" in distinction to "live" or bright timber. Exposure to air darkens all wood; direct sunlight and occasional moistening hasten this change and cause it to penetrate deeper. Prolonged immersion has the same effect, pine wood becoming a dark gray while oak changes to a blackish brown.

Odor, like color, depends on chemical compounds, forming no part of the wood substance itself. Exposure to weather reduces, and often changes the odor, but a piece of dry longleaf pine, cedar, or camphor wood exhales apparently as much odor as ever, when a new surface is exposed.

Heartwood is more odoriferous than sapwood. Many kinds of wood are distinguished by strong and peculiar odors. This is especially the case with camphor, cedar, pine, oak, and mahogany, and the list would comprise every kind of wood in use, were our sense of smell developed in keeping with its importance. Decomposition is usually accompanied by pronounced odors; decaying poplar emits a disagreeable odor, while red oak often becomes fragrant, its smell resembling that of heliotrope.

RESONANCE.

If a log or scantling is struck with the ax or hammer, a sound is emitted which varies in pitch and character with the shape and size of the stick, and also with the kind and condition of wood. Not only can sound be produced by a direct blow, but a thin board may be set vibrating and be made to give a tone by merely producing a suitable tone in its vicinity. The vibrations of the air, caused by the motion of the strings of the piano, communicate themselves to the board, which vibrates in the same intervals as the string and reenforces the note. The note which a given piece of wood may emit varies in pitch directly with the elasticity, and indirectly with the weight, of the wood. The ability of a properly shaped sounding board to respond freely to all the notes within the range of an instrument, as well as to reflect the character of the notes thus emitted (i. e., whether melodious or not), depends, first, on the structure of the wood and next on the uniformity of the same throughout the board. In the manufacture of musical instruments all wood containing defects, knots, cross grain, resinous tracts, alternations of wide and narrow rings, and all wood in which summer and spring wood are strongly contrasted in structure and variable in their proportions, is rejected, and only radial sections (quarter sawed, or split) of wood of uniform structure and growth are used.

The irregularity in structure, due to the presence of relatively large pores and pith rays, excludes almost all our broad-leaved woods from such use, while the number of eligible woods among conifers is limited by the necessity of combining sufficient strength with uniformity in structure, absence of too pronounced bands of summer wood, and relative freedom from resin.

Spruce is the favored resonance wood; it is used for sounding boards both in pianos and violins, while for the resistant back and sides of the latter, the highly elastic hard maple is used. Preferably resonance wood is not bent to assume the final form; the belly of the violin is shaped from a thicker piece, so that every fiber is in the original as nearly unstrained condition as possible, and therefore free to vibrate. All wood for musical instruments is, of course, well seasoned, the final drying in kiln or warm room being preceded by careful seasoning at ordinary temperatures often for as many as seven years or more. The improvement of violins, not by age but by long usage, is probably due, not only to the adjustment of the numerous component parts to each other, but also to a change in the wood itself; years of vibrating enabling any given part to vibrate much more readily.

FIG. 15.—Cross section of a group of wood fibers.

II.—WEIGHT OF WOOD.

A small cross section of wood, as in fig. 15, dropped into water, sinks, showing that the substance of which wood fiber or wood is built up is heavier than water. By immersing the wood successively in heavier liquids, until we find a liquid in which it does not sink, and comparing the weight of the same with water, we find that wood substance is about 1.6 times as heavy as water, and that this is as true of poplar as of oak or pine.

Separating a single cell, as shown in fig. 16, *a*, drying and then dropping it into water, it floats. The air-filled cell cavity or interior reduces its weight, and, like a corked empty bottle, it weighs less than the water. Soon, however, water soaks into the cell, when it fills up and sinks.

Many such cells grown together, as in a block of wood, sink when all or most of them are filled with water, but will float as long as the majority are empty or only partly filled. This is why a green, sappy pine pole soon sinks in "driving" (floating). Its cells are largely filled before it is thrown in, and but little additional water suffices to make its weight greater than that of the water.

In a good-sized white pine log, composed chiefly of empty cells (heartwood), the water requires a very long time to fill up the cells (five years would not suffice to fill them all), and therefore the log may float for many months. When the wall of the wood fiber is very thick (five-eighths or more of the volume), as in fig. 16, *b*, the fiber sinks whether empty or filled. This applies to most of the fibers of the dark summerwood bands in pines, and to the compact fibers of oak or hickory, and many, especially tropical woods, have such thick-walled cells and so little empty or air space that they never float.

Here, then, are the two main factors of weight in wood: The amount of cell wall, or wood substance, constant for any given piece, and the amount of water contained in the wood, variable even in the standing tree, and only in part eliminated in drying.

The weight of the green wood of any species varies chiefly as the second factor, and is entirely misleading if the relative weight of different kinds is sought. Thus some green sticks of the otherwise lighter cypress and gum sink more readily than fresh oak.

The weight of sapwood, or the sappy peripheral part of our common lumber woods, is always great, whether cut in winter or summer. It rarely falls much below 45 pounds and commonly exceeds 55 pounds to the cubic foot, even in our lighter wooded species.

FIG. 16.—Isolated fibers.

It follows that the green wood of a sapling is heavier than that of an old tree, the fresh wood from a disk of the upper part of a tree often heavier than that of the lower part, and the wood near the bark heavier than that nearer the pith, and also that the advantage of drying the wood before shipping is most important in sappy and light kinds.

When kiln dried, the misleading moisture factor of weight is uniformly reduced and a fair comparison possible. For the sake of convenience in comparison the weight of wood is expressed either as the weight per cubic foot, or, what is still more convenient, as specific weight or density. If an old longleaf pine is cut up as shown in fig. 17, the wood of disk No. 1 is heavier than that of disk No. 2, the latter heavier than that of disk No. 3, and the wood of the top disk is found to be only about three-fourths as heavy as that of disk No. 1.

Similarly, if disk No. 2 is cut up as in the figure, the specific weight of the different pieces is:

a about 0.52
b about 0.64
c about 0.67
d, e, f about 0.65

showing that in this disk, at least, the wood formed during the many years' growth, represented in piece a, is much lighter than that of former years. It also shows that the best wood is the middle part, with its large proportion of dark summerwood bands.

Cutting up all disks in the same way, it will be found that the piece a of the first disk is heavier than piece a of the fifth, and that piece c of the first disk excels the piece c of all the other disks. This shows that the wood grown during the same number of years is lighter in the upper parts of the stem; and if the disks are smoothed on their radial surfaces and set up one on top of the other in their regular order for sake of comparison, this decrease in weight will be seen to be accompanied by a decrease in the amount of summer wood. The color effect of the upper disks is conspicuously lighter.

If our old pine had been cut one hundred and fifty years ago, before the outer, lighter wood was laid on, it is evident that the weight of the wood of any one disk would have been found to increase from the center outward, and no subsequent decrease could have been observed.

In a thrifty young pine, then, the wood is heavier from the center outward, and lighter from below upward; only the wood laid on in old age falls in weight below the average.

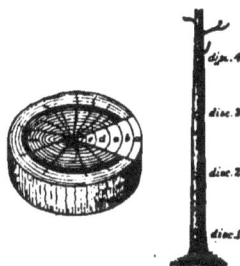

Fig. 17.—Orientation of wood samples.

The number of brownish bands of summer wood are a direct indication of these differences.

If an old oak is cut up in the same manner, the butt cut is also found heaviest and the top lightest, but, unlike the disk of pine, the disk of oak has its firmest wood at the center and each successive piece from the center outward is lighter than its inner neighbor.

Examining the pieces, this difference is not as readily explained by the appearance of each piece as in the case of pine wood. Nevertheless, one conspicuous point appears at once, the pores, so very distinct in oak, are very minute in the wood near the center and thus the wood is far less porous. Studying different trees it is found that, in the pines, wood with narrow rings is just as heavy as, and often heavier than the wood with wider rings, but if the rings are unusually narrow in any part of the disk the wood has a lighter color; that is, there is less summer wood and therefore less weight.

In oak, ash, or elm trees of thrifty growth, the rings fairly wide (not less than one-twelfth inch), always form the heaviest wood, while any piece with very narrow rings is light. On the other hand, the weight of a piece of hard maple or birch is quite independent of the width of its rings.

The bases of limbs (knots) are usually heavy, very heavy in conifers, and also the wood which surrounds them, but generally the wood of the limbs is lighter than that of the stem, and the wood of the roots is the lightest.

In general, it may be said that none of the native woods in common use in this country are, when dry, as heavy as water, i. e., 62 pounds to the cubic foot. Few exceed 50 pounds, while most of them fall below 40 pounds, and much of the pine and other coniferous wood weighs less than 30 pounds per cubic foot.

The weight of the wood is, in itself, an important quality. Weight assists in distinguishing maple from poplar. Lightness, coupled with great strength and stiffness, recommends wood for a thousand different uses. To a large extent weight predicates the strength of the wood, at least in the same species, so that a heavy piece of oak will exceed in strength a light piece of the same species, and in pine it appears probable that, weight for weight, the strength of the wood of various pines is nearly equal.

Weight of kiln-dried wood of different species.

	Specific weight.	Approximate. Weight of—	
		1 cubic foot.	1,000 feet of lumber.
(a) Very heavy woods:		*Pounds.*	*Pounds.*
Hickory, oak, persimmon, osage orange, black locust, hackberry, blue beech, best of elm, and ash	0.70–0.80	42–48	3,700
(b) Heavy woods:			
Ash, elm, cherry, birch, maple, beech, walnut, sour gum, coffee tree, honey locust, best of Southern pine, and tamarack	.60–.70	36–42	3,200
(c) Woods of medium weight:			
Southern pine, pitch pine, tamarack, Douglas spruce, western hemlock, sweet gum, soft maple, sycamore, sassafras, mulberry, light grades of birch and cherry	.50–.60	30–36	2,700
(d) Light woods:			
Norway and bull pine, red cedar, cypress, hemlock, the heavier spruce and fir, redwood, basswood, chestnut, butternut, tulip, catalpa, buckeye, heavier grades of poplar	.40–.50	24–30	2,200
(e) Very light woods:			
White pine, spruce, fir, white cedar, poplar	.30–.40	18–24	1,800

For scientific names see list, p. 72.

Since ordinary lumber contains knots and also more water than is here assumed, and also since its dimensions either exceed or fall short of perfect measurement, the figures in the table are only approximate. Thus, 1,000 feet, B. M., of longleaf pine weighs:

Pounds.

Rough and green 4,500
Boards, rough but seasoned 3,500
Boards, dressed and seasoned 3,000
Flooring, matched, dressed and seasoned 2,500
Weatherboarding beveled and dressed 1,500

III.—MOISTURE IN WOOD.

Water may occur in wood in three conditions: (1) It forms the greater part (over 90 per cent) of the protoplasmic contents of the living cells; (2) it saturates the walls of all cells, and (3) it entirely or at least partly fills the cavities of the lifeless cells, fibers, and vessels. In the sapwood of pine it occurs in all three forms; in the heartwood only in the second form, it merely saturates the walls. Of 100 pounds of water associated with 100 pounds of dry wood substance in 200 pounds of fresh sapwood of white pine, about 35 pounds are needed to saturate the cell walls, less than 5 pounds are contained in living cells, and the remaining 60 pounds partly fill the cavities of the wood fibers. This latter forms the sap as ordinarily understood. It is water brought from the soil, containing small quantities of mineral salts, and in certain species (maple, birch, etc.) it also contains at certain times a small percentage of sugar and other organic matter. These organic substances are the dissolved reserve food, stored during winter in the pith rays, etc., of the wood and bark; generally but a mere trace of them is to be found. From this it appears that the solids contained in the sap, such as albumen, gum, sugar, etc., can not exercise the influence on the strength of the wood which is so commonly claimed for them.

The wood next to the bark contains the most water. In the species which do not form heartwood the decrease toward the pith is gradual, but where this is formed, the change from a more moist to a drier condition is usually quite abrupt at the sapwood limit. In longleaf pine, the wood of the outer 1 inch of a disk may contain 50 per cent of water, that of the next, or second inch, only 35 per cent, and that of the heartwood only 20 per cent. In such a tree the amount of water in any one section varies with the amount of sapwood, and is therefore greater for the upper than the lower cuts, greater for limbs than stems, and greatest of all in the roots.

Different trees, even of the same kind and from the same place, differ as to the amount of water they contain. A thrifty tree contains more water than a stunted one, and a young tree more than an old one, while the wood of all trees varies in its moisture relations with the season of the year.

Contrary to the general belief a tree contains about as much water in winter as in summer. The fact that the bark peels easily in the spring depends on the presence of incomplete, soft tissue found between wood and bark during this season and has little to do with the total amount of water contained in the wood of the stem.

Even in the living tree a flow of sap from a cut occurs only in certain kinds of trees and under special circumstances; from boards, timber, etc., the water does not flow out, as is sometimes believed, but must be evaporated.[1]

---

[1] The seeming exceptions to this rule are mostly referable to two causes, namely: (a) Clefts or "shakes" will allow water contained in them to flow out. (b) From sound wood, if very sappy, water is forced out whenever the wood is warmed, just as water flows from green wood in the stove.

The rapidity with which water is evaporated, that is, the rate of drying, depends on the size and shape of the piece and on the structure of the wood. An inch board dries more than four times as fast as a 4-inch plank and more than twenty times as fast as a 10-inch timber. White pine dries faster than oak. A very moist piece of pine or oak will, during one hour, lose more than four times as much water per square inch from the cross section, but only one-half as much from the tangential, as from the radial section.

In a long timber, where the end or cross sections form but a small part of the drying surface, this difference is not so evident. Nevertheless, the ends dry and shrink first, and being opposed in this shrinking by the more moist adjoining parts, they check, the cracks largely disappearing as seasoning progresses.

High temperatures are very effective in evaporating the water from wood, no matter how humid the air. A fresh piece of sapwood may lose weight in boiling water, and can be dried to quite an extent in hot steam.

Kept on a shelf in an ordinary dwelling wood still retains 8 to 10 per cent of its weight of water, and always contains more water per pound than the surrounding air. Nor is this amount of water constant; the weight of a pan full of shavings varies with the time of day, being on a summer day greatest in the morning and least in the afternoon.

Desiccating the air with chemicals will cause the wood to dry, but wood thus dried at 80° F. will still lose water in the kiln. Wood dried at 120° F. loses water still if dried at 200° F., and this again will lose more water if the temperature is raised. So that absolutely dry wood can not be obtained, and chemical destruction sets in before all the water is driven off.

On removal from the kiln the wood at once takes up water from the air, even in the driest weather. At first the absorption is quite rapid; at the end of a week a short piece of pine, $1\frac{1}{2}$ inches thick, has regained two-thirds of, and, in a few months, all the moisture which it had when air dry, 8 to 10 per cent, and also its former dimensions.

In thin boards all parts soon attain the same degree of dryness; in heavy timbers the interior remains moister for many months, and even years, than the exterior parts. Finally an equilibrium is reached, and then only the outer parts change with the weather.

With kiln-dried wood all parts are equally dry, and when exposed the moisture coming from the air must pass in through the outer parts, and thus the order is reversed. Ordinary timber requires months before it is at its best; kiln-dry timber, if properly handled, is prime at once.

Dry wood, when soaked in water, soon regains its original volume, and in the heartwood portion it may even surpass it; that is to say, swell to a larger dimension than it had when green. With the soaking it continues to increase in weight, the cell cavities filling with water, and if left many months all pieces sink. Yet even after a year's immersion a piece of oak 2 by 2 inches and only 6 inches long still contains air, i. e., it has not taken up all the water it can. By rafting, or prolonged immersion, wood loses some of its weight, soluble materials being leached out, but it is not impaired either as fuel or as building material. Immersion and, still more, boiling and steaming reduce the hygroscopicity of wood and, therefore, also the troublesome "working" or shrinking and swelling.

Exposure in dry air to a temperature of 300° F. for a short time reduces, but does not destroy, the hygroscopicity and with it the tendency to shrink and swell. A piece of red oak, which has been subjected to a temperature of over 300° F., still swells in hot water and shrinks in the kiln.

In artificial drying, temperatures of from 158° F. to 180° F. are usually employed. Pine, spruce, cypress, cedar, etc., are dried fresh from the saw, allowing four days for 1-inch boards; hard woods, especially oak, ash, maple, birch, sycamore, etc., are air-seasoned for three to six months, to allow the first shrinkage to take place more gradually, and are then exposed to the above temperatures in the kiln for about six to ten days for 1-inch lumber. Freshly cut poplar and cottonwood are often dried directly in kilns.

By employing lower temperatures, 100° to 120° F., green oak, ash, etc., can be seasoned in dry kilns without danger to the material. Steaming the lumber is commonly resorted to in order to prevent checking and "casehardening," but not, as has frequently been asserted, to enable the board to dry. Yard-dried lumber is not dry, and its moisture is too unevenly distributed to insure good behavior after manufacture. Careful piling of the lumber, both in the yard and kiln, is essential to good drying. Piling boards on edge or standing them on end is believed to hasten drying. This is true only because in either case the air can circulate more freely around them than when they are piled in the ordinary way. Boards on end dry unequally; the upper half dries much faster than the lower half and horizontal piling is, therefore, preferable.

Since the proportion of sap and heart wood varies with size, age, species, and individual, the following figures must be regarded as mere approximations:

Pounds of water lost in drying 100 pounds of green wood in the kiln.

	Sapwood or outer part.	Heartwood or interior.
(1) Pines, cedars, spruces, and firs	45–65	16–25
(2) Cypress, extremely variable	50–65	18–60
(3) Poplar, cottonwood, basswood	60–65	40–60
(4) Oak, beech, ash, elm, maple, birch, hickory, chestnut, walnut, and sycamore	40–50	30–40

The lighter kinds have the most water in the sapwood, thus sycamore has more than hickory.

IV.—SHRINKAGE OF WOOD.

When a short piece of wood fiber, such as that shown in fig. 18, *A*, is dried it shrinks, its wall grows thinner (as indicated by dotted lines), its width, *a b*, the thickness of the fiber, becomes smaller, and the cavity or opening larger, but, strange to say, the height or length, *b c*, remains the same. In a similar piece of fiber with a thinner wall (fig. 18, *B*) the effect is the same, but the wall being only half as thick the total change is only about half as great.[1]

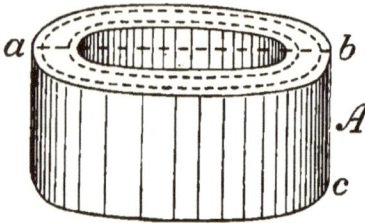

FIG. 18.—Short pieces of wood fibers, one thick, the other thin-walled; magnified.

If sections or pieces of fibers are dried and then placed on moist blotting paper, they will take up water and swell to their original size, though the water has been taken up only by their walls and none has entered into their openings or lumina. This indicates that the water in the cavity or lumen of a fiber has nothing to do with its dimensions, and that if the cell walls are saturated it makes no difference in the volume of a block of pine wood whether the cell cavities are empty as in the heartwood or three-fourths filled as in the sapwood.

FIG. 19.—Isolated cell.

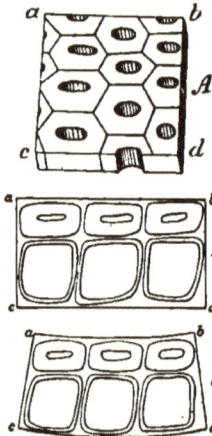

If an entire fiber, as shown in fig. 19, is dried, the wall at its ends *a* and *b*, like those of the sides, grow thinner, and thereby the length of the entire cell grows shorter. Since this length is often a hundred or more times as great as the diameter, the effect of this shrinkage is inappreciable; and if a long board shrinks lengthwise, it is largely due, as we shall see, to quite another cause.

FIG. 20.—Warping of wood.

A thin cross section of several fibers (see fig. 20, *A*) like the piece of a single fiber shrinks when dried, the wall of each fiber becomes thinner, and thus each piece smaller, and the piece on the whole necessarily

---

[1] Though generally true, it must not be supposed that the fibers of all species, or even the fibers of the same tree, shrink exactly in proportion to the thickness of their walls.

shares this diminution of size, the distances, *a b* and *c d*, each becoming shorter. Where the cells are very similar in size and in the thickness of their walls, as in the case of piece *A*, fig. 20, *a b* and *c d* become shorter by about the same amount; but if the piece is made up of fibers, some of which have thin and others thick walls, as piece *B*, fig. 20, then the row of thick-walled cells shrinking much more than the row of thin-walled cells, the piece becomes unevenly shrunk or warped as shown in fig. 20, *C*. Not only is the piece warped, but the force which led to this warping continues to strain the interior parts of the piece in different directions.

Since in all our woods cells with thick walls and cells with thin walls are more or less intermixed, and especially as the spring wood and summer wood nearly always differ from each other in this respect, strains and tendencies to warp are always active when wood dries out, because the summer wood shrinks more than the spring wood, heavier wood in general more than light wood of the same kind.

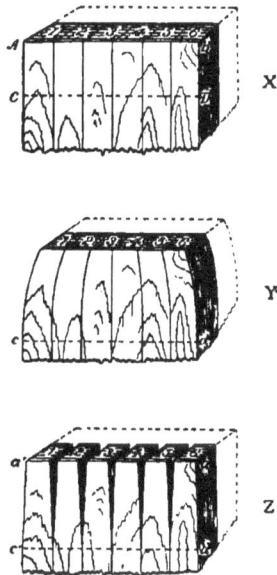

If the piece *A*, fig. 20, after drying, is placed edgewise on moist blotting paper, the cells on the underside, at *c d*, take up moisture from the paper and swell before the upper cells at *a b* receive any moisture. This causes the underside of the piece to become longer than the upper side and, as in the case of piece *C*, warping occurs. Soon, however, the moisture penetrates to all the cells and the piece straightens out. A thin board behaves exactly like this minute piece, only the process is slower and more easily observed. But while a thin board of pine curves laterally, it remains quite straight lengthwise, since in this direction both shrinkage and swelling are small. A thin disk or cross

Fig. 21.—Formation of checks.

section swells, and when moistened on one side warps as readily in one direction as in another. If a green board is exposed to the sun with one side, warping is produced by removal of water and consequent shrinkage of the upper side, and the course of the process is simply reversed.

As already stated, wood loses water faster from the end than from the longitudinal faces. Hence the ends shrink at a different rate from the interior parts.

In a timber, the width *A B* (fig. 21, *X*) may have shortened (fig. 21, *Y*), while a short distance from the end *c d*, the original width is still preserved. This should produce a bending of the parts toward the center of the piece as shown in exaggeration at *Y*, but the rigidity of

the several parts of the timber prevents such bending and the consequent strain leads to their separation as shown at Z, the end surface
of the timber being "checked."

As the timber dries out, the line c d becomes shorter, the parts 1 to 6
are allowed to approach again, and the checks close up and are no
longer visible.

The faster the drying at the surface, the greater is the difference in
the moisture of the different parts, and hence the greater the strains
and consequently also the amount of checking. This becomes very
evident when fresh wood is placed in the sun, and still more in a hot kiln.
While most of these smaller checks are thus only temporary, closing
up again, some large radial checks remain and even grow larger as
drying progresses. Their cause is a different one and will presently be
explained.

The temporary checks not only occur at the ends, but are developed
on the sides also, only to a much smaller
degree. They become especially annoying on the surface of thick planks
of hard woods, and also on peeled logs
when exposed to the sun.

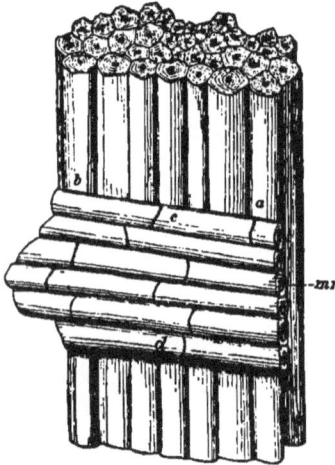

So far we have considered the wood
as if made up only of parallel fibers all
placed longitudinally in the log. This,
however, is not the case. A large part
of the wood is formed by the medullary or pith rays. In pine over 15,000
of these occur on a square inch of a
tangential section, and even in oak the
very large rays, which are readily visible to the eye, represent scarcely a
hundredth part of the number which
the microscope reveals.

As seen in fig. 22 the cells of these
rays have their length at right angles
to the direction of the wood fibers.

FIG. 22.—Small pith ray in oak. *a, b,* wood
fibers; *c, d,* cells of pith ray.

If a large pith ray of white oak is whittled out and allowed to dry it
is found to shrink greatly in the direction from c to d (fig. 22), while, as
we have stated, the fibers to which the ray is firmly grown in the wood
do not shrink in the same direction. Therefore, in the wood, as the
cells of the pith ray dry, they pull on the longitudinal fibers and try
to shorten them, and, being opposed by the rigidity of the fibers, the
pith ray is greatly strained. But this is not the only strain it has to
bear. Since the fibers from a to b (fig. 22) shrink as much again as the
pith ray in this, its longitudinal direction, the fibers tend to shorten
the ray, and the latter, in opposing this, prevents the former from

shrinking as much as they otherwise would. Thus the structure is subjected to two severe strains at right angles to each other, and herein lies the greatest difficulty of wood seasoning, for whenever the wood dries rapidly these fibers have not the chance to "give" or accommodate themselves, and hence fibers and pith rays separate and checks result which, whether visible or not, are detrimental in the use of the wood.

The contraction of the pith rays parallel to the length of the board is probably one of the causes of the small amount of longitudinal shrinkage which has been observed in boards.[1] The smaller shrinkage of the pith rays along the radius of the log (the length of the pith ray) opposing the shrinkage of the fibers in this direction becomes one of the causes of the second great trouble in wood seasoning, namely, the difference in the amount of the shrinkage along the radius and that along the rings or tangent.

This greater tangential shrinkage appears to be due, in part, to the cause just mentioned, but also to the fact that the greatly shrinking bands of summer wood are interrupted, along the radius, by as many bands of porous spring wood, while they are continuous in the tangential direction. In this direction, therefore, each such band tends to shrink, as if the entire piece were composed of summer wood, and since the summer wood represents the greater part of the wood substance, this tendency of greater tangential shrinkage prevails.

The effect of this greater tangential shrinkage affects every phase of woodworking. It leads to permanent checks, and causes the log to split open on drying.

Sawed in two, the flat sides of the log become convex, as in fig. 23; sawed into a timber, it checks along the median line of the four faces, and if converted

Fig. 23.—Effects of shrinkage.

into boards, the latter take on the forms shown in fig. 23, all owing to the greater tangential shrinkage of the wood.

Briefly, then, shrinkage of wood is due to the fact that the cell walls grow thinner on drying. The thicker cell walls and therefore the heavier wood shrinks most, while the water in the cell cavities does not influence the volume of the wood. Owing to the great difference of cells in shape, size, and thickness of walls, and still more in their arrangement, shrinkage is not uniform in any kind of wood. This irregularity produces strains, which grow with the difference between

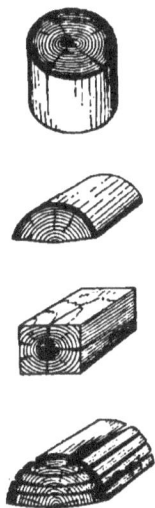

---

[1] In addition to this all fibers having an oblique position, as those at pith rays and knots, also the oblique, tapering ends of all fibers contribute to this longitudinal shrinkage, since one component of their normal shrinkage is longitudinal.

adjoining cells and are greatest at the pith rays. These strains cause warping and checking, but exist even where no outward signs are visible; they are greater if the wood is dried rapidly than if dried slowly, but can never be entirely avoided.

Temporary checks are caused by the more rapid drying of the outer parts of any stick; permanent checks are due to the greater shrinkage, tangentially, along. the rings than that along the radius. This, too, is the cause of most of the ordinary phenomena of shrinkage, such as the difference in behavior of entire and quartered logs "bastard" (tangent) and "rift" (radial) boards, etc., and explains many of the phenomena erroneously attributed to the influence of bark, or of the greater shrinkage of outer and inner parts of any log.

Once dry, wood may be swelled again to its original size by soaking in water, boiling, or steaming. Soaked pieces, on drying, shrink again as before; boiled and steamed pieces do the same, but to a slightly less degree. Neither hygroscopicity, i. e., the capacity of taking up water, nor shrinkage of wood can be overcome by drying at temperatures below 200° F. Higher temperatures, however, reduce these qualities, but nothing short of a coaling heat robs wood of the capacity to shrink and swell. Rapidly dried in the kiln, the wood of oak and other hard woods "case-harden," that is, the outer part dries and shrinks before the interior has a chance to do the same, and thus forms a firm shell or case of shrunken, commonly checked wood around the interior. This shell does not prevent the interior from drying, but when this drying occurs, the interior is commonly checked along the medullary rays, as shown in fig. 24. In practice this occurrence can be prevented by steaming the lumber in the kiln, and still better by drying the wood in the open air or in a shed before placing in the kiln. Since only the first shrinking is apt to check the wood, any kind of lumber which has once been air dried (three to six months for 1-inch stuff) may be subjected to kiln heat without any danger. Kept in a bent or warped condition during the first shrinking, the wood retains the shape to which it was bent and firmly opposes any attempt at subsequent straightening.

Sapwood, as a rule, shrinks more than heartwood of the same weight, but very heavy heartwood may shrink more than lighter sapwood. The amount of water in wood is no criterion of its shrinkage, since in wet wood most of the water is held in the cavities, where it has no effect on the volume.

The wood of pine, spruce, cypress, etc., with its very regular structure, dries and shrinks evenly and suffers much less in seasoning than the wood of broad-leafed trees. Among the latter, oak is the most difficult to dry without injury. Small-sized split ware and "rift" boards season better than ordinary boards and planks.

FIG. 24.—"Honeycombed" board. The checks or cracks form along the pith rays.

To avoid "working" or warping and checking, all high-grade stock is carefully seasoned, preferably in a kiln, before manufacture. Thicker pieces may be made of several parts glued together; larger surfaces are made in panels or of smaller pieces covered with veneer. Boring is sometimes resorted to to prevent the checking of wooden columns. Since repeated swelling increases the injuries due to seasoning, wood should be protected against moisture when once it is dry.

Since the shrinkage of our woods has never been carefully studied, and since wood, even from the same tree, varies within considerable limits, the figures given in the following table are to be regarded as mere approximations. The shrinkage along the radius and that along the tangent (parallel to the rings) are not stated separately in the following table, and the figures represent an average of the shrinkage in the two directions. Thus, if the shrinkage of soft pine is given at 3 inches per hundred, it means that the sum of radial and tangential shrinkage is about 6 inches, of which about 4 inches fall to the tangent and 2 inches to the radius, the ratio between these varying from 3 to 2, a ratio which practically prevails in most of our woods.

Since only an insignificant longitudinal shrinkage takes place (being commonly less than 0.1 inch per hundred), the change in volume during drying is about equal to the sum of the radial and tangential shrinkage, or twice the amount of linear shrinkage indicated in the table.

Thus, if the linear average shrinkage of soft pine is 3 inches per hundred, the shrinkage in volume is about 6 cubic inches for each 100 cubic inches of fresh wood.

Approximate shrinkage of a board, or set of boards, 100 inches wide, drying in the open air.

	Shrink-age.
	Inches.
(1) All light conifers (soft pine, spruce, cedar, cypress)	3
(2) Heavy conifers (hard pine, tamarack, yew), honey locust, box elder, wood of old oaks	4
(3) Ash, elm, walnut, poplar, maple, beech, sycamore, cherry, black locust	5
(4) Basswood, birch, chestnut, horse chestnut, blue beech, young locust	6
(5) Hickory, young oak, especially red oak	Up to 10

V.—MECHANICAL PROPERTIES OF WOOD.

Every joist and studding, every rafter, sash, and door, the chair we sit on, the floor we walk on, the wood of the wagon or boat we ride in are all continually tested as to their stiffness and strength, their hardness and toughness. Every step from the simple splitting of a shingle or stave to the construction of the most elegant carriage or sideboard involves a knowledge, not only of one, but of several, of the mechanical properties of the material.

In the shop the fitness of the wood for a given purpose never depends on any one quality alone, but invariably upon a combination of several qualities. A spoke must not only be strong, it must be stiff to hold its shape, it must be tough to avoid shattering to pieces, and it must also be hard or else its tenons will become loose in their mortises.

Selecting wood in this way, the woodworker has learned almost all that is at present known about his material, but in many cases the great difficulty which always attends the judgment of complex phenomena has led to erroneous conclusions, and not a few well-established beliefs have their origin more in accidental error of observation than in fact.

The experimenter endeavors to avoid this complexity by testing the wood for each kind of resistance separately; when tested as to their stiffness, the pieces are all shaped, placed, and loaded alike. The wood is selected with a definite object in view; it is green or dry, clear or knotty, straight or crossgrained, according as he wishes to find out the influence of each of these conditions. If pine and oak are to be compared, the pieces are from the same position in the tree and are tried under exactly the same conditions, and thus the case is simplified.

But even results thus arrived at can not be used indiscriminately, and the figures on the strength of oak given in any book must not be supposed to apply to all oak, if tested in the given manner. This is due to the fact that a piece of wood is not simply a material but a structure, just as much as' a railroad bridge or a balloon frame, and as such varies greatly even in the wood of the same tree, nay, more than that, even in the same year's growth of the same cross section of a log.

A scantling resists bending; it is stiff. On removal of the load it straightens; it is elastic. A column, a prop, or the spoke of a wagon wheel resists being crushed endwise. So does the upper side of a joist or beam when loaded, while the underside of the beam or of an ax handle suffers in tension. The tenons of a window sash or door tend to break out their mortises, the wood has to resist shearing along the fibers; the steel edge of the eye tends to cut into the hammer handle, it tries to shear it across the grain, and every nail, screw, bore hole, or mortise tends to split the board and tries the wood as to its cleavability, while all "bent" ware, from the wicker basket to the one-piece felly or ship's knee, involves its flexibility.

STIFFNESS.

If 100 pounds placed in the middle of a stick 2 by 2 inches and 4 feet long, supported at both ends, bend or "deflect" this stick one-eighth of an inch (in the middle), then 200 pounds will bend it about one-fourth inch, 300 pounds three-eighths inch, the deflection varying directly as the load. Soon, however, a point is reached where an additional 100 pounds adds more than one-eighth inch to the deflection—the limit of elasticity has been reached. Taking another piece from the straight grained and perfectly clear plank of the same depth and width, but 8 feet long, the load of 100 pounds will cause it to bend not only one-eighth inch, but will deflect it by about 1 inch. Doubling the length reduces the stiffness eightfold. Stiffness then decreases as the cube of the length.

Cutting out a piece 2 by 4 inches and 4 feet long, placing it flatwise so that it is double the width of the former stick and loading it with 100 pounds, we find it bending only one-sixteenth inch; doubling the width doubles the stiffness.

Setting the same 2 by 4 inch piece on edge, so that it is 2 inches wide and 4 inches deep, the load of 100 pounds bends it only about one sixty-fourth inch; doubling the thickness increases the stiffness about eightfold.

It follows that if we double the length and wish to retain the same stiffness we must also double the thickness of the piece.

A piece of wood is usually stiffer with the annual rings set vertically than if the rings are placed horizontally to the load.

Crossgrained and knotty wood, to be sure, is not as stiff as clear lumber; a knot on the upper side of a joist, which must resist in compression, is, however, not so detrimental as a knot on the lower side, where it is tried in tension.

Every large timber which comes from the central part of the tree contains knots, and much of its wood is

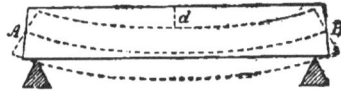

FIG. 25.—Bending a beam.

cut more or less obliquely across the grain, both conditions rendering such material comparatively less stiff than small clear pieces.

The same stick of pine, green or wet, is only about two-thirds as stiff as when dry. A heavy piece of longleaf pine is stiffer than a light piece; heavy pine in general is stiffer than light pine, but a piece of hickory, although heavier than the pine, may not be as stiff as the piece of longleaf pine, and a good piece of larch exceeds in stiffness any oak of the same weight.

In the same tree stiffness varies with the weight, the heavier wood being the stiffer; thus the heavier wood of the butt log is stiffer than that of the top; timber with much of the heavy summer wood is stiffer than timber of the same kind with less summer wood. In old trees (of pine) the center of the tree and the sap are the least stiff; in thrifty young pine the center is the least stiff, but in young second growth hard woods it is the stiffest.

Since it is desirable, and for many purposes essential, to know before hand that a given piece with a given load will bend only a given amount, the stiffness of wood is usually stated in a uniform manner and under the term "modulus (measure) of elasticity."

If AB, fig. 25, is a piece of wood, and d the deflection produced by a weight or load, the elasticity of the wood, as usually stated, is found by the formula:

$$\text{Modulus of elasticity} = \frac{W\ l^3}{4\ D\ bd^3}$$

where W is the weight, l the length, b and d the breadth and depth of the stick, and D the deflection for the load W. In the following table the woods are grouped according to their stiffness. The figures are only rough approximations which are based on the data given in Vol. IX of the Tenth Census. The first column contains the above modulus, the second shows how many pounds will produce a deflection of 1 inch in a stick 1 by 1 by 12 inches, assuming that it could endure such bending within the limits of elasticity, and the third column gives the number of pounds which will bend a stick 2 by 2 inches and 10 feet long through 1 inch.

The stick is assumed to rest on both ends; if it is a cantilever, i. e., fastened at one end and loaded at the other, it bears but half as much load at its end for the same deflection.

From the third column it is easy to find how many pounds would bend a piece of the same kind of other dimensions. A 2 by 4 inch bears eight, a 2 by 6 inch twenty-seven times as much as the 2 by 2 inch; a piece 8 feet long is about twice as stiff as a 10-foot piece; a piece 12 feet, only about three-fifths, 14 feet one-third, 16 feet two-ninths, 18 feet one-sixth, and 20 feet one-eighth as stiff.

The number of pounds which will bend any piece of sawed timber by 1 inch may be found by using the formula:

$$\text{Necessary weight} = \frac{4\,\text{E}\,bd^3}{l^3}$$

where E is the figure in the first column, b, d, l, breadth, depth, and length of the timber in inches. If the deflection is not to exceed one-half inch, only one-half the load, and if one-fourth inch, only one-fourth the load, is permissible.

To allow for normal irregularities in the structure of wood itself, as well as in the aggregate structure of timbers, an allowance is made on the numbers which have been found by experiment; this allowance is called the "factor of safety." Where the selection of the wood is not very perfect, the load is a variable one, and the safety of human life depends on the structure, the factor is usually taken quite high, as much as 6 or 10, i. e., only one-sixth or one-tenth of the figures given in the tables is considered safe, and the beam is made six to ten times as heavy as the calculation requires.

Table of stiffness (modulus of elasticity) of dry wood.—General averages.

Species.	Modulus of elasticity $E = \frac{W l^3}{4 D b d^3}$ per square inch.	Approximate weight which deflects by 1 inch a piece—	
		1 by 1 inch and 12 inches long.	2 by 2 inches and 10 feet long.
	Pounds.	*Pounds.*	*Pounds.*
(1) Live oak, good tamarack, longleaf, Cuban, and short-leaf pine, good Douglas spruce, western hemlock, yellow and cherry birch, hard maple, beech, locust, and the best of oak and hickory....................	1, 680, 000	3. 900	62
(2) Birch, common oak, hickory, white and black spruce, loblolly and red pine, cypress, best of ash, elm, and poplar and black walnut	1, 400, 000	3, 200	51
(3) Maples, cherry, ash, elm, sycamore, sweet gum, butternut, poplar, basswood, white, sugar and bull pine, cedars, scrub pine, hemlock, and fir..................	1, 100, 000	2, 500	40
(4) Box elder, horse chestnut, a number of western soft pines, inferior grades of hard woods..................	1, 100, 000	¹ 2, 500	40

¹ Less than.

CROSS-BREAKING OR BENDING STRENGTH.

When the addition of 100 pounds to the load on our 2 by 2 inch piece begins to add more than one-eighth inch to the deflection, that is, when the stick has been bent beyond its "elastic limit," it still requires an increase of 30 to 50 per cent to the load before the stick breaks. The load which is borne before the limit of elasticity is reached indicates the strength of the wood up to this important point; the load which causes it to break represents its absolute strength, or the "cross-breaking or bending strength" as it is commonly called.

In longleaf pine the former (modulus of strength at the elastic limit),[1] is commonly about three-fourths of the latter. If left loaded for a considerable time, a load but little greater than that which brings the stick to its elastic limit will cause it to break, and this load should therefore not be exceeded.

Unlike the stiffness, the strength of a timber varies approximately with the squares of the thickness and decreases directly with increasing length and not with the cube of this latter dimension. Thus, if our piece 2 by 2 inches and 4 feet long can bear 1,000 pounds before it breaks, a 2 by 4 inch laid flat will break with about 2,000 pounds, and if set edgewise, it requires about 4,000 pounds to break it, while a piece of the same kind of 2 by 2 inches, and double the length (8 feet), breaks with half the original load, or only 500 pounds.

All conditions of the material which influence the stiffness also influence the bending strength. Seasoning increases, moisture decreases, the strength; knots and crossgrain depress it and both are more dangerous on the lower than on the upper side. But while the conifers with their simple cell structure excel in stiffness, the better hard woods

[1] The elastic limit in this case is somewhat of an arbitrary quantity, namely, the point where 100 pounds produces a deflection 50 per cent greater than the preceding 100 pounds.

develop the greater strength in bending. Like elasticity and stiffness, the strength is expressed in a uniform manner by the so-called "modulus of rupture," to permit ready estimation of the strength of any given piece. This modulus refers to the resistance which the parts most strained, "the extreme fiber," offer. For reasons above stated, in practice a factor of safety is employed. as in all these calculations of resistance. The figures usually tabulated are obtained by the formula:

$$\text{Strength of extreme fiber} = \frac{3\,W\,l}{2bd^2}$$

where W is the breaking load, l the length, b and d the breadth and depth of the tested piece of wood.

The following table presents our common woods grouped as to their strength in bending. The load, as before, is supposed to act altogether in the middle. Column 1 gives the strength of the extreme fiber, as explained above; column 2, the number of pounds which will break a piece 1 by 1 inch and 12 inches long, and column 3, the strength of a stick 2 by 2 inches and 10 feet long, from which the strength of any given piece can readily be estimated, allowing, however, for defects, which increase with the size. Thus, if a good piece of pine 2 by 2 inches and 10 feet long breaks with 400 pounds, a 2 by 4 inch set on edge requires 1,600 pounds, a 2 by 6 inch, 3,600 pounds, a 2 by 8 inch piece 6,400 pounds to break it. If a piece 2 by 4 inches and 10 feet long breaks with 1,000 pounds, a 2 by 4 inch and 12 feet long piece breaks with about 1,300 pounds, one 16 feet with 1,000 pounds, etc., and if a factor of safety of 10 is allowed, only one-tenth of the above loads are permissible.

A board one-half inch by 12 inches and 10 feet long contains as much wood as a 2 by 3 inch of the same length, and if placed edgewise should offer four times as much resistance to breaking. Owing to its small breadth, however, it "twists" when loaded, and in most cases, therefore, bears less than the 2 by 3 inch. To prevent this twisting, joists are braced, and the depth of timbers is made not to exceed four times their thickness.

Short deep pieces shear out or split before their strength in bending can fully be called into play.

Strength in cross-breaking of well-seasoned, select pieces.

	Strength of the extreme fiber $f = \dfrac{3\,Wl}{2\,bd^2}$ per square inch.	Approximate weight which breaks a stick—	
		1 by 1 inch and 12 inches long.	2 by 2 inches and 10 feet long.
(1) Robinia (locust), hard maple, hickory, oak, birch, best ash and elm, longleaf, shortleaf, and Cuban pines, tamarack..	*Pounds.* 13,000	*Pounds.* 720	*Pounds.* 570
(2) Soft maple, cherry, ash, elm, walnut, inferior oak, and birch, best poplar, Norway, loblolly and pitch pines, black and white spruce, hemlock and good cedar......	10,000	550	440
(3) Tulip, basswood, sycamore, butternut, poplars, white and other soft pines, firs, and cedars.................	6,500	350	280

TENSION AND COMPRESSION.

When a piece of wood is pulled lengthwise, in the manner shown in fig. 26, part of the fibers are torn asunder or broken, but many are merely pulled or shredded out from between their neighbors. Since failure in tension thus involves lateral adhesion as well as strength of fibers, it is affected not only by the nature and dimensions of the fibers but also by their arrangement. Owing to their transverse position the medullary rays (a large part of all woods) offer but one-tenth to one-twentieth as much resistance as the main body of fibers and moreover weaken the timber by disturbing the straight course of the fibers and the regularity of the entire structure.

The resistance is also much affected by the position of the grain. The perfectly cross-grained piece *a* (fig. 27) sustains but about one-tenth to one-twentieth of the load which is supported by the straight-grained piece *c*,

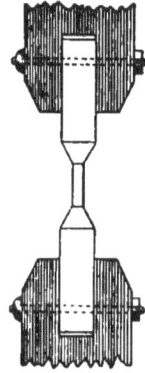

FIG. 26.—Specimen in tension test.

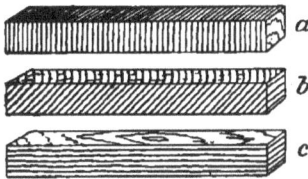

FIG. 27.—Straight and cross grained wood.

and it is evident that the piece *b*, which represents the ordinary case of crossgrain, is likewise weakened by the oblique position of the grain.

This explains the detrimental influence of a knot on the underside of a board, as in fig. 28. Since the lower side of the board, in bending, is stretched, the upper side being compressed, the fibers of the lower side

FIG. 28.—Effect of knots and their position.

are subjected to tension and the wood of the knot, like the piece of crossgrained wood, offers but little resistance. Commonly the defect is greatly increased by a season check in the knot itself, so that the knot affects the strength of the board like a saw cut of equal depth.

Tested in compression endwise (fig. 29), the fibers act as so many hollow columns firmly grown together, and when the load becomes too great the piece fails in the manner illustrated in fig. 31.

This failure is a very complex phenomenon; in wood like pine the fibers of the plain in which failure occurs become separated into small bodies; they tear apart and cease to behave as one solid body but act as a large number of very small independent pieces. Like the strands of a rope these small bodies offer but little resistance to compression; they bend over, and the piece "buckles."

FIG. 29.—Compression endwise.

It is evident that a vertical position and a regular arrangement of the fibers increase the resistance, and that therefore the medullary rays and oblique position of fibers in crossgrained and knotty timber tend to reduce the strength in compression.

FIG. 30.—Longitudinal shearing.

From the following table of strength in tension and compression it will be seen that these two are not always proportional, the stiffer conifers excelling in the latter, the tougher hard woods in the former:

Ratio of strength in tension and compression, showing the difference between rigid conifers and tough hard woods.

	Ratio: Tensile strength. $R = $ compressive strength.	A stick 1 square inch in cross section. Weight required to—	
		Pull apart.	Crush endwise.
		Pounds.	*Pounds.*
Hickory	3.7	32,000	8,500
Elm	3.8	29,000	7,500
Larch	2.3	19,400	8,000
Longleaf pine	2.2	17,300	7,400

Strength in compression of common American woods in well-seasoned select pieces.

[Approximate weight per square inch of cross section requisite to crush a piece of wood endwise.]

Pounds.

(1) Black locust, yellow and cherry birch, hard maple, best hickory, longleaf and Cuban pines, and tamarack .... 9,000+

(2) Common hickory, oak, birch, soft maple, walnut, good elm, best ash, shortleaf and loblolly pines, western hemlock, and Douglas fir .... 7,000+

(3) Ash, sycamore, beech, inferior oak, Pacific white cedar, canoe cedar, Lawson's cypress, common red cedar, cypress, Norway and superior spruces, and fir .... 6,000+

(4) Tulip, basswood, butternut, chestnut, good poplar, white and other common soft pines, hemlock, spruce, and fir .... 5,000+

(5) Soft poplar, white cedar, and some western soft pines, and firs .... 4,000+

SHEARING.

When, in a structure like that shown in fig. 30, a weight is placed on J and the tenon T by downward pressure breaks out the piece $A\ B\ C\ D$,

FIG. 31.—Various forms of failure. A and B, compression endwise; C, shearing (the bolt of a stirrup passed through the mortise and sheared out the end); D, tension. The lower figure indicates the number of pounds per square inch which produced the failure in tests by the Division of Forestry. No. 116 (upper figure on each piece) is white pine; Nos. 1, 2, and 5 are longleaf pine, about one-fifth natural size.

this is said to shear out along the fiber. In the same manner, if the shoulder *A B C D* in fig. 30, is pushed off along *B D*, it is sheared, and if *B D* and *C E* are each 1 inch, the surface thus sheared off is 1 square inch, and the weight necessary to do this represents the shearing strength per square inch of the particular kind of wood. This resistance is small when compared to that of tension and compression.

In general wet or green wood shears about one-third more easily than dry wood; a surface parallel to the rings (tangent) shears more easily than one parallel to the medullary rays. The lighter conifers and hard woods offer less resistance than the heavier kinds, but the best of pine shears one-third to one-half more readily than oak or hickory, indicating that great shearing strength is characteristic of "tough" woods.

Resistance to shearing along the fiber.

	Per square inch.
	Pounds.
(1) Locust, oak, hickory, elm, maple, ash, birch	[1] 1,000
(2) Sycamore, longleaf, Cuban, and shortleaf pine, and tamarack	600
(3) Tulip, basswood, better class of poplar, Norway, loblolly and white pine, spruce, red cedar.	400
(4) Softer poplar, hemlock, white cedar, fir	[2] 400

[1] Over.   [2] Less than.

NOTE.—Resistance to shearing, although a most important quality in wood, has not been satisfactorily studied. The values in the above table, taken from various authors, lack a reliable experimental basis and can be considered as only a little better than guesswork.

INFLUENCE OF WEIGHT AND MOISTURE ON STRENGTH.

It has been stated that heavy wood is stronger than lighter wood of the same kind, and that seasoning increases all forms of resistance. Let us examine why this is so.

Since the weight of dry wood depends on the number of fibers and the thickness of their walls, there must be more fibers per square inch of cross section in the heavy than in the light piece of the same kind,[1] and it is but natural that the greater number of fibers should also offer greater resistance, i. e., have the greater strength.

The beneficial influence of drying and consequent shrinking is two-fold: (1) In dry wood a greater number of fibers occur per square inch, and (2) the wood substance itself, i. e., the cell walls, become firmer. A piece of green longleaf pine, 1 by 1 inch and 2 inches long, is only about 0.94 by 0.96 inch and 2 inches long when dry; its cross section is 10 per cent smaller than before, but it still contains the same number of fibers. A dry piece 1 by 1 inch, therefore, contains 10 per cent more fibers than a green piece of the same size, and it is but fair to suppose that its resistance or strength is also about 10 per cent greater.

The influence of the second factor, though unquestionably the more important one, is less readily measured. In 100 cubic inches of wood

[1] This imperfect assumption is used only for comparison.

substance the material of the cell walls takes up about 50 cubic inches of water and thereby swells up, becoming about 150 cubic inches in volume. In keeping with this swelling the substance becomes softer and less resistant. In pine wood this diminution of resistance, according to experiments, seems to be about 50 per cent, and the strength of the substance therefore is inversely as the degree of saturation or solution.

HARDNESS AND SHEARING ACROSS THE GRAIN.

When the solid steel plunger P in fig. 32 descends on the piece of wood w, the first effect is to press it into the wood of the upper surface without affecting the interior or lower part. The wood is thus tried with regard to its hardness. If a perforated steel plate is substituted for the solid plate the effect of the plunger is at first the same, but soon the fibers some distance from the steel are seen to bend, and finally the piece of wood fails in shearing across the grain. Hardness and shearing across the grain are closely related. The

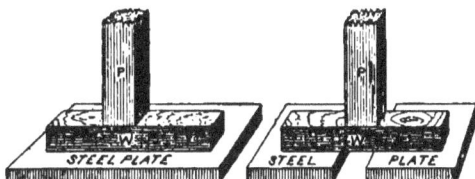

Fig. 32.—Test in hardness and shearing across the grain.

former is the more important quality, however, since abrasion and indentation, the two failures in hardness, are the common cause of loosening of tenons in the mortise, of the handle in the ax, etc.

Heavy wood is harder than lighter wood; the wood of the butt, therefore, is harder than that of the top; the darker summer wood harder than the light-colored spring wood. Moisture softens, and seasoning, therefore, hardens wood.

Placing the rings vertical helps the wood to resist indentation. Though harder wood resists saw and chisel more than softer wood, the working quality of the wood is not always a safe criterion of its hardness.

The following indicates the hardness of our common woods:

1. Very hard woods requiring over 3,200 pounds per square inch to produce an indentation of one-twentieth inch: Hickory, hard maple, osage orange, black locust, persimmon, and the best of oak, elm, and hackberry.

2. Hard woods requiring over 2,400 pounds per square inch to produce an indentation of one-twentieth inch: Oak, elm, ash, cherry, birch, black walnut, beech, blue beech, mulberry, soft maple, holly, sour gum, honey locust, coffee tree, and sycamore.

3. Middling hard woods, requiring over 1,600 pounds per square inch to produce an indentation of one-twentieth inch: The better qualities of Southern and Western hard pine, tamarack and Douglas spruce, sweet gum, and the lighter qualities of birch.

4. Soft woods requiring less than 1,600 pounds per square inch to produce an indentation of one-twentieth inch: The greater mass of coniferous wood; pine, spruce, fir, hemlock, cedar, cypress, and redwood; poplar, tulip, basswood, butternut, chestnut, buckeye, and catalpa.

CLEAVABILITY.

When an ax is struck into a piece of wood as shown in fig. 33 the cleft projects beyond the blade of the ax and the process is not one of cutting, but of tension across the grain. The ax presses on a lever, $a\ b$, while the surface in which the transverse tension takes place is reduced almost to a line across the stick at b. If the wood is very elastic, the cleft runs far ahead of the ax, the lever arm $a\ b$ is long, and the resistance to splitting proportionately small. Elasticity, therefore, helps splitting, while great shearing strength, a good measure for transverse tension and hardness hinder it.

FIG. 33.—Cleavage.

Wood splits naturally along two normal planes, the most readily along the radius, because the arrangement of fibers and pith rays is radial, and next along the tangent, or with the annual rings, because the softer spring wood forms continuous planes in this direction. Cleavage along the radius, however, is from 50 to 100 per cent easier, and only in case of cross grain, etc., the cleavage along the ring becomes the easier. In the wood of conifers, wood fibers and pith rays are very regular, the former in perfect radial series or rows, and cleavage is, therefore, very easy in this direction. The same is brought about in the oak by the very high pith rays, but where they are thick and low, as in sycamore, and generally in the butt cuts and about knots, they impede cleavage by causing a greater irregularity in the course of the wood fibers. The greater the contrast of spring and summer wood, the easier the cleavage tangentially or in the direction of the rings. This is especially marked in conifers and also in woods like oak, ash, and elm, where the spring wood appears as a continuous series of large pores. Very slow growth influences tangential cleavage, narrow-ringed oak breaks out and splits less regularly even in a radial direction; in conifers, however, this difference scarcely exists. Weight of wood affects the cleavage but little; in heavy wood the entrance of the ax, to be sure, is resisted with more force, but the greater elasticity of the wood, on the other hand, counterbalances this resistance. Irregularities in the course of the fibers, whether spiral growth, crossgrain, or in form of knots, all aid in resisting cleavage. Knotty bolts are split more easily from the upper end, since the cleft then runs around the knots (see p. 23). Moisture softens the wood and reduces lateral adhesion, and therefore wood splits more easily when green than when dry.

FLEXIBILITY.

Pine is brittle, hickory is flexible; the former breaks, the latter bends. Being the opposite of stiffness, want of stiffness would seem to indicate flexibility. This, however, is only partly true; hickory and ash are stiff and yet among the most flexible of woods. Their small dimensions cause shavings and thin strands of most woods to appear pliable. For this reason the pliable, twisted wicker willow is not a fair measure of the flexibility of the wood of this species. Generally hard woods are more flexible than conifers, wood of the butt surpassing in this respect that of the main part of the stem, the latter being usually superior to that of the limbs. Moisture softens wood and thereby increases its flexibility. Knots and crossgrain diminish flexibility, but the irregular structure of elm, ash, etc. (particularly the arrangement of bodies of extremely firm fibers, like so many strands, among the softer tissue, as well as the interlacement of fibers, due to post-cambial growth), favorably influences the flexibility of these woods.

TOUGHNESS.

So far the load by which the exhibition of the various kinds of strength in compression, tension, cross bending, etc., was produced has always been assumed as applied slowly and gradually. When a wagon goes lumbering along a cobble pavement the load on the spokes is not thus applied. Every stone deals the wheel a blow, and a mile's journey means many thousand blows to every wheel rim and spoke. In chopping, the ax handle is jarred and a handle made of pine wood, which shears easily along the fiber, would soon be shattered to pieces. Loads thus applied are "shocks," and resistance to this form of loading requires a combination of various kinds of strength possessed only by "tough" woods. Toughness is a familiar word to woodworkers, and yet is rarely defined. Tough wood must be both strong and pliable. Thus a willow is not tough when dry; it is weak and brittle, and requires, notwithstanding its small lateral dimensions, to be moistened and twisted or sheared into still smaller strands so that its fibers are subjected almost exclusively to tension, if great deflection and great strength are to be combined (handles of wicker baskets). Hickory is both strong and pliable; in the dimensions of a willow twig it can be used almost like a rope. The term "tough," therefore, is properly applied to woods like hickory and elm and improperly to willow.

Judging from the behavior of elm and hickory, wood may be pronounced "tough" if it offers great resistance to—

(1) Longitudinal shearing over 1,000 pounds per square inch,
(2) Tension over 16,000 pounds per square inch,

and permits, when tested dry, of an aggregate distortion in compression and tension amounting to not less than 3 per cent.

For instance, of a piece of dry hickory (*H. alba*) we may expect—

Strength in shearing...........................pounds..   1,200
Strength in tension.................................do.... 25,000

Distortion in tension..........................per cent..   2.03
Distortion in compression..........................do....   1.55

Total distortion..................................do....   3.58

PRACTICAL CONCLUSIONS.

From the foregoing considerations a few valuable facts, mostly familiar to the thoughtful woodworker, may be deduced:

In *framing*, where light and stiff timber is wanted, the conifers excel; where heavy but steady loads are to be supported, the heavier conifers, hard pine, spruce, Douglas spruce, etc., answer as well as hard woods, which are costlier and heavier for the same amount of stiffness. On the other hand, if small dimensions must be used, and especially if moving loads are to be sustained, hard woods are safest, and in all cases where the load is applied in form of "shocks" or jars, only the tougher hard woods should be employed. The heavier wood surpasses the lighter of the same species in all kinds of strength, so that the weight of dry wood and the structural features indicative of weight may be used as safe signs in selecting timber for strength.

In *shaping* wood it is better, though more wasteful, to split than to saw, because it insures straight grain and enables a more perfect seasoning.

For *sawed stock* the method of "rift" or "quarter" sawing, which has so rapidly gained favor during the last decade, deserves every encouragement. It permits of better selection and of more advantageous disposition of the wood; rift-sawed lumber is stronger, wears better, seasons well, and is least subject to "working" or warping.

All hardwood material which *checks or warps* badly during seasoning should be reduced to the smallest practicable size before drying, to avoid the injuries involved in this process; and wood once seasoned should never again be exposed to the weather, since all injuries due to seasoning are thereby aggravated. Seasoning increases the strength of wood in every respect, and it is therefore of great importance to protect wooden structures, bearing heavy weights, against moisture.

Knots, like crossgrain and other defects, reduce the strength of timber. Where choice exists, the knotty side of the joist should be placed uppermost, i. e., should be used in compression.

Season checks in timber are always a source of weakness; they are more injurious on the vertical than on the horizontal faces of a stringer or joist, and their effect continues even when they have closed up, as many do, and are no longer visible.

Rafted timber, kiln-dried or steamed lumber are, as far as our present knowledge extends, as strong as other kinds, and wherever any of these processes aids in a more uniform or perfect seasoning, it increases the strength of the material.

Pine "bled" for turpentine is as strong as "unbled."

Time of felling, whether season of the year or phase of the moon, does not influence strength, except that summer-felled hard wood rarely seasons as perfectly as that felled in the fall, and to this extent an indirect influence may be observed, as well as by the fact that fungi and insects have a better opportunity for developing.

Warm countries and sunny exposures generally produce heavier and stronger timber, and conditions favorable to the growth of the species also improve its quality. But exceptions occur; neither fast nor slow growth is an infallible sign of strong wood, and it is the character of the annual ring, rather than its width, and particularly the proportion of summer wood, which determines the quality of the material.

VI.—CHEMICAL PROPERTIES OF WOOD.

Wood dried at 300° F. is composed of over 99 per cent of organic and less than 1 per cent of inorganic matter; the latter remains as ashes when wood is burned.

Wood consists of a skeleton of cellulose, permeated by a mixture of other organic substances, collectively designated by the name of lignin, and particles of mineral matter or ashes.

Cellulose is the common substance of which plant cells form their cases or walls; in flax, the entire fiber is almost pure cellulose, but the amount of cellulose obtained from wood, by the common processes, rarely exceeds one-half of its dry weight. Cellulose is identical in composition with starch, but unlike the latter it resists alcoholic fermentation, though the plants themselves, as well as decay-producing fungi, are able to reconvert it into starch, from which it seems originally derived, and also to change it into various forms of sugar.[1] Lignin is as yet a chemical puzzle. The substances forming it are carbohydrates like cellulose itself, but of slightly different proportions and distinguished by greater solubility in acids, and by other chemical properties.

In 100 pounds of wood (dried at 300° F.) and of cellulose the following proportions are found:

	Wood.	Cellulose.
	Pounds.	*Pounds.*
Carbon	49	44.4
Hydrogen	6	6.1
Oxygen	44	49.3

[1] Chemists have succeeded in producing reconversion into grape sugar, and though the methods thus far employed are expensive, it is to be expected that in the near future wood will become the principal source of both vinegar and alcohol.

This composition of wood is fairly uniform for different species.

At ordinary temperatures wood is a very stable compound; both in air and under water it remains the same for centuries, and only when living organisms attack it with their strong solvents and convertants do change and decay set in.

Heated to 300° F. wood gives off only water, though some slight chemical changes are noticeable even at this temperature. If the heat is increased, gases of pungent odor and taste are evolved, and if the temperature is sufficiently raised, the gases are ignited, forming the flame of the fire, while the remaining solid part glows like an ignited charcoal, giving much heat, but no flame. The amount of heat produced by wood varies. If first dried at 300° F., 100 pounds of poplar wood should give as much heat as 100 pounds of hickory. In the natural state, however, this is not the case.

The beneficial effect of thorough seasoning for firewood appears from the following consideration:

One hundred pounds of wood as sold in the wood yards contains in round numbers 25 pounds of water, 74 pounds of wood, and 1 pound of ashes.

The 74 pounds of wood are composed of 37 pounds of carbon, 4.4 pounds of hydrogen, and 32 pounds of oxygen.

In burning (which is a process of oxidation) 4 pounds of hydrogen are already combined with 32 pounds of oxygen and there are only the 37 pounds of carbon and 0.4 pounds of hydrogen available in heat production. Thus only about one-half the weight of the wood substance itself is heat producing while every pound of water combined in the wood requires about 600 units of heat to evaporate it, and thus diminishes the value of the wood as fuel. Hence under the most favorable circumstances 100 pounds of green wood (50 per cent moisture) furnishes about 150,000 units [1] of heat; 100 pounds of half dry (30 per cent moisture) about 230,000 units; 100 pounds of air dry (20 per cent moisture) about 280,000 units; 100 hundred pounds of air dry (10 per cent moisture) about 320,000 units; 100 pounds of kiln-dry (2 per cent moisture) about 350,000 units.

In the ordinary stove or other small apparatus the evil effect of moisture in the wood is very much increased since combustion is materially interfered with.

One hundred pounds of ordinary charcoal furnishes 700,000 units of heat but the same quantity of charcoal produced at a temperature of 2,000° F. furnishes nearly 800,000 units of heat.

Conifers and the lighter hard woods produce more flame, while the heavy hard woods furnish a good bed of live coal and exceed the former by 25 to 30 per cent in production of heat with ordinary appliances.

---

[1] A unit of heat in this case is the amount of heat which raises the temperature of 1 pound of water by 1.8° F. or 1° C.

Heated in a closed chamber or covered with earth, as in charcoal pits, the wood is prevented from burning and a variety of changes occur, depending on the rate of heating. If the temperature is raised gradually so that the wood is heated several hours before a temperature of 600° F. is reached the process is called dry distillation. In this process the wood is destroyed. It forms at first "red" or "brown" coal, still resembling wood, and finally charcoal proper. This coal is darker, heavier, conducts heat and electricity better, requires a greater heat to ignite, and produces more heat in burning the higher the temperature under which it is formed.

One hundred pounds of wood (dried at 300° F.) leaves only about 30 pounds of charcoal. In common practice much less charcoal (18 to 20 per cent) is produced. In this change from wood to coal the volume is diminished by about one-half, so that a cord of wood which contains about 100 cubic feet of wood solid would be converted into 50 cubic feet at best.

Of the 70 pounds of gaseous products which 100 pounds of wood lose, during coaling, in being heated up to 700° F., about 63 pounds become volatile before the temperature of 550° F. is reached.

If condensed in a cooler, about three-fourths of the 63 pounds of volatile matter first evolved is found to be wood-vinegar, from which about 4 pounds of pure acetic acid, the only source of perfectly pure vinegar, is obtained. Besides acetic acid, the liquid contains wood spirits and a quantity of various allied substances.

After the first stage of dry distillation, a large part of the products developed can not be liquefied in the ordinary cooler. They are gases like the illuminating gas, mostly belonging to the marsh gas series; they lack oxygen and thus show that the available oxygen has been nearly exhausted in the preceding part of the process. Products of the later stages are tars and heavy oils, volatile only at high temperatures. Here also belong the substances known collectively as wood creosote, employed as antiseptics in wood impregnation.

Warmed in dilute nitric acid with a little chlorate of potash, the cells of a piece of wood may be separated, each cell remains intact, but its wall is reduced in thickness and material; the lignin substances being dissolved out, only the cellulose is left. In commercial cellulose manufacture, soda, sulphates, and of late chiefly sulphites are substituted for the nitric acid. The wood is chipped, boiled in the respective solution under high pressures, the residue is washed, and the remaining cellulose bleached and ready for use. As a matter of economy the residual liquid is evaporated and the soda used over again.

When resinous wood, "fat pine," "lightwood," such as the knots and stumps of longleaf, pitch, and other pines, is heated in a kiln or retort, the resins ooze out, are collected, and in distillation with steam yield turpentine and rosin. The resins and their components vary with the species; the balsam of fir is limpid, its turpentine remains clear on exposure; the resin of pines is very viscid, their turpentines readily oxidize and darken when brought in contact with air. Resins are gathered more commonly either from cracks, such as "wind" and "ring shakes," as in the case of larch and fir (Venetian turpentine), or else from wounds made especially for this purpose, as in the case of naval stores gathered from pines. This latter process is known as "bleeding," "tapping," or "orcharding," and is at present the principal method of obtaining turpentines and rosins.

On burning resinous wood, wood tar, etc., in a smoldering fire, soot is deposited on the walls and partitions of the specially constructed soot pit. It is then collected, but must be freed of various products of dry distillation, by carefully heating to red heat before it becomes the lampblack used in printers' ink and otherwise much employed in the arts.

Many kinds of wood and the bark of most trees contain tannin. To serve in tanning the bark must contain at least 3 per cent of tannin; the kinds mostly used vary from 5 to 15 per cent, and even the best probably never furnish over 20 per cent in the average. The use of tan bark involves considerable disadvantages. It is difficult to dry and preserve, very liable to mold, bulky, and therefore expensive to ship and store, and very variable in the amount of tannin which it contains.

To avoid these difficulties the tannic compounds are, in recent times, leached out of the finely ground bark and wood, condensed by evaporation, and shipped as extracts containing 80 to 90 per cent of tannin.

The manufacture of pulp as well as the production of fiber capable of being spun and woven, are also technological uses of wood, which rely partly upon chemical reactions.

VII.—DURABILITY AND DECAY.

All wood is equally durable under certain conditions. Kept dry or submerged, it lasts indefinitely. Pieces of pine have been unearthed in Illinois which have lain buried 60 or more feet deep for many centuries. Deposits of sound logs of oak, buried for unknown ages, have been unearthed in Bavaria; parts of the piles of the lake dwellers, driven more than two thousand years ago, are still intact.

On the radial section of a piece of pine timber, with one of the shelf-like, fungus growths, as shown in fig. 34, both bark and wood are seen to be affected. A small particle of the half-decayed wood presents pictures like that of fig. 35. Slender, branching threads are seen to attach themselves closely to the walls of the cells, and to pierce these in all directions. Thus these little threads of fungus mycelium soon form a perfect network in the wood, and as they increase in number they dissolve the walls, and convert the wood substance and cell contents into sugar-like food for their own consumption. In some cases it is the woody cell wall alone that is attacked. In other cases they confine themselves to eating up the starch found in the cells, as shown in fig. 36, and merely leave a stain (bluing of lumber). In all cases of decay we find the vegetative bodies, these slender threads of fungi, responsible for the mischief. These fine threads are the vegetative body of the fungus, the little shelf is its fruiting body, on which it produces myriads of little spores (the seeds of fungi). Some fungi attack only conifers, others hard woods; many are confined to one species of tree and perhaps no one attacks all kinds of wood. One kind produces "red rot," others "bluing." In one case the decayed tracts are tubular, and in the direction of the fibers the wood is "peggy." In other cases no particular shapes are discernible.

FIG. 34.—"Shelf" fungus on the stem of a pine. (Hartig.) a, sound wood; b, resinous "light" wood; c, partly decayed wood or punk; d, layer of living spore tubes; e, old filled up spore tubes; f, fluted upper surface of the fruiting body of the fungus, which gets its food through a great number of fine threads (the mycelium), its vegetative tissue penetrating the wood and causing its decay.

Cutting off a disk of loblolly pine, washing it, and then laying it in a clean, shady place in the sawmill, its sapwood will be found stained in a few days. Nor is this mischief confined to the surface; it penetrates the sapwood of the entire disk. From this it appears that the spores must have been in the air about the mill, and also that their germination and the growth of the threads or mycelium is exceedingly rapid. (Watching the progress of mold on a piece of bread teaches the same thing.) Placing a fresh piece of sapwood on ice, another into a dry kiln, and soaking a few others in solutions of corrosive sublimate (mercuric chloride) and other similar salts, we learn that the fungus growth is retarded by cold, prevented and killed by temperatures over 150° F., and that salts of mercury, etc., have the same effect. The fact that seasoned pieces if exposed are not so readily attacked by fungi shows that the moisture in air-dry wood is insufficient for fungus growth.

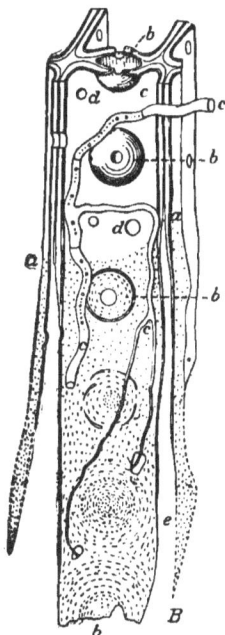

FIG. 35.—Fungus threads in pine wood. (Hartig.) a, cell wall of the wood fibers; b, bordered pits of these fibers; c, thread of mycelium of the fungus; d, holes in the cell walls made by the fungus threads, which gradually dissolve the walls as shown at e, and thus break down the wood structure.

From this it appears that warmth, preferably between 60° and 100° F., combined with abundance of moisture (but not immersion), is the most important condition favoring decay, and that the defense lies in the proper regulation or avoidance of these

conditions, or else in the use of poisonous salts, which prevent the propagation of fungi.

It is also apparent, therefore, why wood decays faster in Alabama than in Wisconsin, faster in the swamps than on the plains, and why the presence of large quantities of decaying wood about the yard, constantly producing fresh supplies of spores, stimulates decay. Covering with tar or impregnating with creosote, salts of mercury, copper, etc., enables even sapwood to last under the most trying conditions. Contact with the ground assures most favorable moisture conditions for fungus growth, and the higher temperatures near the surface of the ground, together with the ever-present supply of spores, cause rot in a post to start at the surface more readily than 30 inches below.

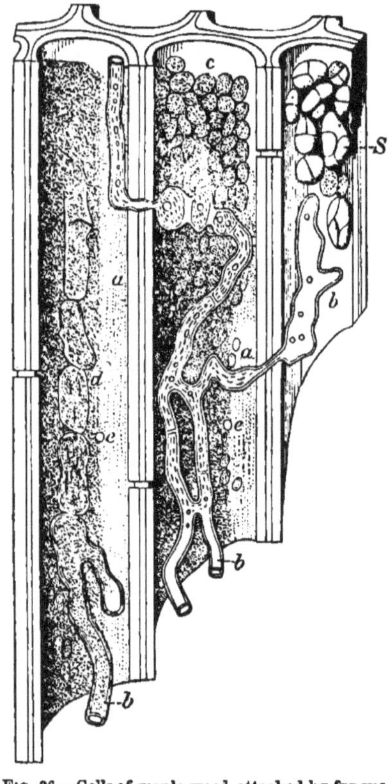

The use of means to prevent decay is therefore desirable where timber is placed in positions favorable to fungus growth, as in railway ties; and all joists and timber in contact with damp brick walls, as also all building material whose perfect seasoning is prevented by the absence of proper circulation of air, should be specially protected. In the former cases it is economy to apply preservative processes; in the latter a sanitary necessity. Wood covered with paint, etc., before it is perfectly seasoned, falls a prey to "dry rot;" the fungus finds abundance of moisture, and the protection intended for the wood protects its enemy, the fungus. Since charcoal resists the solvents of fungi, charring the outer parts of posts makes, if well done, namely, so as not to open checks into the interior of the wood, a very fine protection.

FIG. 36.—Cells of maple wood attacked by fungus threads (*Nectria cinnabarina* Mayer). Section of three wood fibers showing the threads of the fungus branching in their cavities and consuming the starch stored in these cells. *a*, interior or cavity of cells; *b*, threads of the fungus; *c*, partly destroyed starch grains; *d*, dead portions of the fungus thread together with débris; *e*, holes bored by the fungus through the cell walls; *S*, starch grains just being attacked.

Under ordinary circumstances, only the second great factor of decay, i. e., the moisture condition, can be controlled.

Perfect seasoning, preferably kiln-drying, before using, and protection against the entrance of moisture by tar, paints, and other covers, when put in place, prolong the life of wooden structures. Where such a covering is too expensive, good ventilation at least is necessary. Contact surfaces, where timber rests on timber or brick, should in all cases be especially protected.

Different species differ in their resistance to decay. Cedar is more durable than pine and oak better than beech, but in most cases the conditions of warmth and moisture in particular locations have so much to do with durability that often an oak post outlasts one of cedar, even in the same line of fence, and predictions of durability become mere guesswork.

Containing more ready-made food, and in forms acceptable to a great number of different kinds of fungi, the sapwood is more subject to decay than the heartwood, doubly so where the latter is protected by resinous substances, as in pine and cedar. Several months of immersion improves the durability of sapwood, but only impregnation with preservative salts seems to render it perfectly secure. Once attacked by fungi, wood becomes predisposed to further decay.

Wood cut in the fall is more durable than that cut in summer, only because the low temperature of the winter season prevents the attack of the fungi, and the wood is thus given a fair chance to dry. Usually summer-felled wood, on account of prevalent high temperature and exposure to sun, checks more than winter-felled wood, and since all season checks favor the entrance of both moisture and fungus, they facilitate destruction. Where summer-felled wood is worked up at once and protected by kiln-drying no difference exists. The phases of the moon have no influence whatever on durability.

In sawing timber much of the wood is bastard cut; at these places water enters much more readily, and for this reason split and hewn timber and ties generally resist decay perhaps better than if sawed.

The attacks of beetles, as well as those of the shipworm, can not here be considered; like chisel or saw they are mechanical injuries against which none of our woods are proof.

Range of durability in railroad ties.

	Years.		Years.
White oak and chestnut oak	8	Redwood	12
Chestnut	8	Cypress and red cedar	10
Black locust	10	Tamarack	7 to 8
Cherry, black walnut, locust	7	Longleaf pine	6
Elm	6 to 7	Hemlock	4 to 6
Red and black oaks	4 to 5	Spruce	5
Ash, beech, maple	4		

The durability of wood, exposed to the changes of the weather, and where painting, after thorough seasoning, is impracticable, is increased by impregnating it with various salts or other chemicals, which prevent the fungus from feeding on the wood. The wood is first steamed, to open the pores and remove the hardened surface coating of sap and dirt, and a liquid solution of the preservative material is then injected with the assistance of heat and pressure.

The most efficient fluids used on a large scale are bichloride of zinc and creosote, or both combined. The "life" of railroad ties is thereby increased to twice and three times its natural duration.

HOW TO DISTINGUISH THE DIFFERENT KINDS OF WOOD.

By B. E. FERNOW and FILIBERT ROTH.

The carpenter or other artisan who handles different woods becomes
familiar with those he employs frequently, and learns to distinguish
them through this familiarity, without usually being able to state the
points of distinction. If a wood comes before him with which he is
not familiar, he has, of course, no means of determining what it is,
and it is possible to select pieces even of those with which he is well
acquainted, different in appearance from the general run, that will make
him doubtful as to their identification. Furthermore, he may distinguish between hard and soft pines, between oak and ash, or between
maple and birch, which are characteristically different; but when it
comes to distinguishing between the several species of pine or oak or
ash or birch, the absence of readily recognizable characters is such that
but few practitioners can be relied upon to do it. Hence, in the market
we find many species mixed and sold indiscriminately.

To identify the different woods it is necessary to have a knowledge
of the definite, invariable differences in their structure, besides that of
the often variable differences in their appearance. These structural
differences may either be readily visible to the naked eye or with a
magnifier, or they may require a microscopical examination. In some
cases such an examination can not be dispensed with, if we would make
absolutely sure. There are instances, as in the pines, where even our
knowledge of the minute anatomical structure is not yet sufficient to
make a sure identification.

In the following key an attempt has been made—the first, so far as we
know, in English literature—to give a synoptical view of the distinctive
features of the commoner woods of the United States, which are found
in the markets or are used in the arts. It will be observed that the
distinction has been carried in most instances no further than to genera
or classes of woods, since the distinction of species can hardly be accom-
plished without elaborate microscopic study, and also that, as far as
possible, reliance has been placed only on such characteristics as can
be distinguished with the naked eye or a simple magnifying glass, in
order to make the key useful to the largest number. Recourse has
also been taken for the same reason to the less reliable and more variable general external appearance, color, taste, smell, weight, etc.

The user of the key must, however, realize that external appearance,
such, for example, as color, is not only very variable but also very difficult to describe, individual observers differing especially in seeing and

59

describing shades of color. The same is true of statements of size, when relative, and not accurately measured, while weight and hard ness can perhaps be more readily approximated. Whether any feature is distinctly or only indistinctly seen will also depend somewhat on individual eyesight, opinion, or practice. In some cases the resemblance of different species is so close that only one other expedient will make distinction possible, namely, a knowledge of the region from which the wood has come. We know, for instance, that no longleaf pine grows in Arkansas and that no white pine can come from Alabama, and we can separate the white cedar, giant arbor vitæ of the West and the arbor vitæ of the Northeast, only by the difference of the locality from which the specimen comes. With all these limitations properly appreci ated, the key will be found helpful toward greater familiarity with the woods which are more commonly met with.

The features which have been utilized in the key and with which— their names as well as their appearance—therefore, the reader must famil iarize himself before attempting to use the key, are mostly described as they appear in cross section. They are:

(1) Sapwood and heartwood (see p. 13), the former being the wood from the outer and the latter from the inner part of the tree. In some

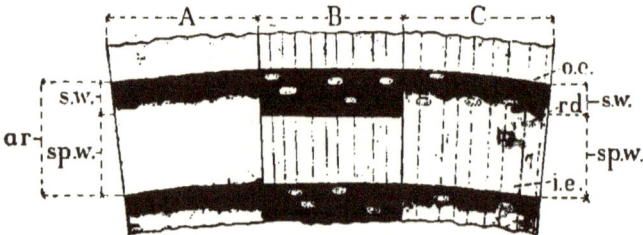

Fig. 37.—" Non-porous " woods. *A*, fir; *B*, "hard" pine; *C*, soft pine; *ar*, annual ring; *o. e.*, outer edge of ring; *i. e*, inner edge of ring; *s. w.*, summer wood; *sp. w*, spring wood; *rd*, resin ducts.

cases they differ only in shade, and in others in kind of color, the heart wood exhibiting either a darker shade or a pronounced color. Since one can not always have the two together, or be certain whether he has sapwood or heartwood, reliance upon this feature is, to be sure, unsat isfactory, yet sometimes it is the only general characteristic that can be relied upon. If further assurance is desired, microscopic structure must be examined; in such cases reference has been made to the pres ence or absence of tracheids in pith rays and the structure of their walls, especially projections and spirals.

(2) Annual rings, their formation having been described on page 14. (See also figs. 37–39.) They are more or less distinctly marked, and by means of such marking a classification of three great groups of wood is possible.

(3) Spring wood and summer wood, the former being the interior (first formed wood of the year), the latter the exterior (last formed) part

of the ring. The proportion of each and the manner in which the one merges into the other are sometimes used, but more frequently the manner in which the pores appear distributed in either.

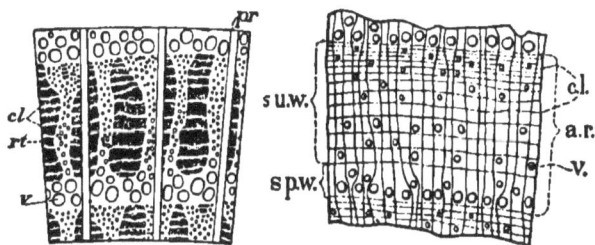

FIG. 38.—"Ring-porous" woods—white oak and hickory. *a. r.*, annual ring; *su. w.*, summer wood; *sp. w.*, spring wood; *v*, vessels or pores; *c. l.*, "concentric" lines; *rt*, darker tracts of hard fibers forming the firm part of oak wood; *pr*, pith rays.

(4) Pores, which are vessels cut through, appearing as holes in cross section, in longitudinal section as channels, scratches, or indentations. (See p. 19 and figs. 38 and 39.) They appear only in the broad-leaved, so called, hard woods; their relative size (large, medium, small, minute, and indistinct, when they cease to be visible individually by the naked eye) and manner of distribution in the ring being of much importance, and especially in the summer wood, where they appear singly, in groups, or short broken lines, in continuous concentric, often wavy, lines, or in radial branching lines.

(5) Resin ducts (see p. 16 and fig. 37), which appear very much like pores in cross section, namely, as holes or lighter or darker colored dots, but much more scattered. They occur only in coniferous woods, and their presence or absence, size, number, and distribution are an important distinction in these woods.

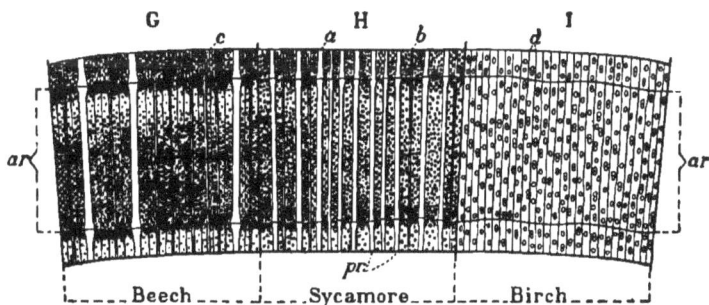

FIG. 39.—"Diffuse-porous" woods. *ar*, annual ring; *pr*, pith rays which are "broad" at *a*, "fine" at *b*, "indistinct" at *d*.

(6) Pith rays (see p. 17 and figs. 38 and 39), which in cross section appear as radial lines, and in radial section as interrupted bands of varying breadth, impart a peculiar luster to that section in some woods. They are most readily visible with the naked eye or with a magnifier in the

broad-leaved woods. In coniferous woods they are usually so fine and closely packed that to the casual observer they do not appear. Their breadth and their greater or less distinctness are used as distinguishing marks, being styled fine, broad, distinct, very distinct, conspicuous, and indistinct when no longer visible by the naked (strong) eye.

(7) Concentric lines, appearing in the summer wood of certain species more or less distinct, resembling distantly the lines of pores but much finer and not consisting of pores. (See fig. 38.)

Of microscopic features, the following only have been referred to:

(8) Tracheids, a description of which is to be found on page 20.

(9) Pits, simple and bordered, especially the number of simple pits in the cells of the pith rays, which lead into each of the adjoining tracheids.

For standards of weight, consult table on page 28; for standards of hardness, table on page 47.

Unless otherwise stated the color refers always to the fresh cross section of a piece of dry wood; sometimes distinct kinds of color, sometimes only shades, and often only general color effects appear.

HOW TO USE THE KEY.

Nobody need expect to be able to use successfully any key for the distinction of woods or of any other class of natural objects without some practice. This is especially true with regard to woods, which are apt to vary much, and when the key is based on such meager general data as the present. The best course to adopt is to supply one's self with a small sample collection of woods, accurately named. Small, polished tablets are of little use for this purpose. The pieces should be large enough, if possible, to include pith and bark, and of sufficient width to permit ready inspection of the cross section. By examining these with the aid of the key, beginning with the better-known woods, one will soon learn to see the features described and to form an idea of the relative standards which the maker of the key had in mind. To aid in this, the accompanying illustrations will be of advantage. When the reader becomes familiar with the key, the work of identifying any given piece will be comparatively easy. The material to be examined must, of course, be suitably prepared. It should be moistened; all cuts should be made with a very sharp knife or razor and be clean and smooth, for a bruised surface reveals but little structure. The most useful cut may be made along one of the edges. Instructive, thin, small sections may be made with a sharp penknife or razor, and when placed on a piece of thin glass, moistened and covered with another piece of glass, they may be examined by holding them toward the light.

Finding, on examination with the magnifier, that it contains pores, we know it is not coniferous or nonporous. Finding no pores collected in the spring-wood portion of the annual ring, but all scattered (diffused) through the ring, we turn at once to the class of "Diffuse-porous woods." We now note the size and manner in which the pores are distributed through the ring. Finding them very small and neither conspicuously grouped, nor larger nor more abundant in the spring wood, we turn to the third group of this class. We now note the pith rays, and finding them neither broad nor conspicuous, but difficult to distinguish, even with the magnifier, we at once exclude the wood from the first two sections of this group and place it in the third, which is represented by only one kind, cottonwood. Finding the wood very soft, white, and on the longitudinal section with a silky luster, we are further assured that our determination is correct. We may now turn to the list of woods and obtain further information regarding the occurrence, qualities, and uses of the wood.

Sometimes our progress is not so easy; we may waver in what group or section to place the wood before us. In such cases we may try each of the doubtful roads until we reach a point where we find ourselves entirely wrong and then return and take up another line; or we may anticipate some of the later-mentioned features and finding them apply to our specimen, gain additional assurance of the direction we ought to travel. Color will often help us to arrive at a speedy decision. In many cases, especially with conifers, which are rather difficult to distinguish, a knowledge of the locality from which the specimen comes is at once decisive. Thus, northern white cedar, and bald cypress, and the cedar of the Pacific will be identified, even without the somewhat indefinite criteria given in the key.

KEY TO THE MORE IMPORTANT WOODS OF NORTH AMERICA.

[The numbers preceding names refer to the List of Woods following the Key.]

I **Non-porous woods.**—Pores not visible or conspicuous on cross section, even with magnifier. Annual rings distinct by denser (dark colored) bands of summer wood (fig. 37).

II. **Ring-porous woods.**—Pores numerous, usually visible on cross section without magnifier. Annual rings distinct by a zone of large pores collected in the spring wood, alternating with the denser summer wood (fig. 38).

III. **Diffuse-porous woods.**—Pores numerous, usually not plainly visible on cross section without magnifier. Annual rings distinct by a fine line of denser summer wood cells, often quite indistinct; pores scattered through annual ring, no zone of collected pores in spring wood (fig. 39).

NOTE.—The above described three groups are exogenous, i. e., they grow by adding annually wood on their circumference. A fourth group is formed by the endogenous woods, like yuccas and palms, which do not grow by such additions.

I.—NON-POROUS WOODS.

(Includes all coniferous woods.)

A. Resin ducts wanting.[1]
  1. No distinct heartwood.
    a. Color effect yellowish white; summer wood darker yellowish (under microscope pith ray without tracheids)..................(Nos. 9–13) FIRS.
    b. Color effect reddish (roseate) (under microscope pith ray with tracheids), (Nos. 14 and 15) HEMLOCK.
  2. Heartwood present, color decidedly different in kind from sapwood.
    a. Heartwood light orange red; sapwood, pale lemon; wood, heavy and hard ....(No. 38) YEW.

ADDITIONAL NOTES FOR DISTINCTIONS IN THE GROUP.

Spruce is hardly distinguishable from fir, except by the existence of the resin ducts, and microscopically by the presence of tracheids in the medullary rays. Spruce may also be confounded with soft pine, except for the heartwood color of the latter and the larger, more frequent, and more readily visible resin ducts.

In the lumber yard, hemlock is usually recognized by color and the slivery character of its surface. Western hemlocks partake of this last character to a less degree.

Microscopically the white pine can be distinguished by having usually only one large pit, while spruce shows three to five very small pits in the parenchyma cells of the pith ray communicating with the tracheid.

The distinction of the pines is possible only by microscopic examination. The following distinctive features may assist in recognizing, when in the log or lumber pile, those usually found in the market:

The light, straw color, combined with great lightness and softness, distinguishes the white pines (white pine and sugar pine) from the hard pines (all others in the market), which may also be recognized by the gradual change of spring wood into summer wood. This change in hard pines is abrupt, making the summer wood appear as a sharply defined and more or less broad band.

---

[1] To discover the resin ducts a very smooth surface is necessary, since resin ducts are frequently seen only with difficulty, appearing on the cross section as fine whiter or darker spots normally scattered singly, rarely in groups, usually in the summer wood of the annual ring. They are often much more easily seen on radial, and still more so on tangential sections, appearing there as fine lines or dots of open structure of different color or as indentations or pin scratches in a longitudinal direction.

b. Heartwood purplish to brownish red; sapwood yellowish white; wood soft to medium hard light, usually with aromatic odor .(No. 6) RED CEDAR.

c. Heartwood maroon to terra cotta or deep brownish red; sapwood light orange to dark amber, very soft and light, no odor; pith rays very distinct, specially pronounced on radial section............(No. 7) REDWOOD.

3. Heartwood present, color only different in shade from sapwood, dingy-yellowish brown.

    a. Odorless and tasteless...............................(No. 8) BALD CYPRESS.

    b. Wood with mild resinous odor, but tasteless....(Nos. 1–4) WHITE CEDAR.

    c. Wood with strong resinous odor and peppery taste when freshly cut..(No. 5) INCENSE CEDAR.

B. Resin ducts present.

1. No distinct heartwood; color white, resin ducts very small, not numerous,
(Nos. 33–36) SPRUCE.

2. Distinct heartwood present.

    a. Resin ducts numerous, evenly scattered through the ring.

        a'. Transition from spring wood to summer wood gradual; annual ring distinguished by a fine line of dense summer-wood cells; color, white to yellowish red; wood soft and light.......(Nos. 18–21) SOFT PINES.[1]

        b'. Transition from spring wood to summer wood more or less abrupt; broad bands of dark-colored summer wood; color from light to deep orange; wood medium hard and heavy ....(Nos. 22–32) HARD PINES.[1]

    b. Resin ducts not numerous nor evenly distributed.

        a'. Color of heartwood orange-reddish, sapwood yellowish (same as hard pine); resin ducts frequently combined in groups of 8 to 30, forming lines on the cross section (tracheids with spirals),
(No. 37) DOUGLAS SPRUCE.

        b'. Color of heartwood light russet brown; of sapwood yellowish brown; resin ducts very few, irregularly scattered (tracheids without spirals)............................... (Nos. 16 and 17) TAMARACK.

The Norway pine, which may be confounded with the shortleaf pine, can be distinguished by being much lighter and softer. It may also, but more rarely, be confounded with heavier white pine, but for the sharper definition of the annual ring, weight, and hardness.

The longleaf pine is strikingly heavy, hard, and resinous, and usually very regular and narrow ringed, showing little sapwood, and differing in this respect from the shortleaf pine and loblolly pine, which usually have wider rings and more sapwood, the latter excelling in that respect.

The following convenient and useful classification of pines into four groups, proposed by Dr. H. Mayr, is based on the appearance of the pith ray as seen in a radial section of the spring wood of any ring:

Section I. Walls of the tracheids of the pith ray with dentate projections.

    a. One to two large, simple pits to each tracheid on the radial walls of the cells of the pith ray.—Group 1. Represented in this country only by *P. resinosa.*

    b. Three to six simple pits to each tracheid, on the walls of the cells of the pith ray.—Group 2. *P. taeda, palustris,* etc., including most of our "hard" and "yellow" pines.

Section II. Walls of tracheids of pith ray smooth, without dentate projections.

    a. One or two large pits to each tracheid on the radial walls of each cell of the pith ray.—Group 3. *P. strobus, lambertiana,* and other true white pines.

    b. Three to six small pits on the radial walls of each cell of the pith ray. Group 4. *P. parryana,* and other nut pines, including also *P. balfouriana.*

[1] Soft and hard pines are arbitrary distinctions and the two not distinguishable at the limit.

II.—RING-POROUS WOODS.

[Some of Group D and cedar elm imperfectly ring-porous.]

A. Pores in the summer wood minute, scattered singly or in groups, or in short broken lines, the course of which is never radial.
  1. Pith rays minute, scarcely distinct.
    a. Wood heavy and hard; pores in the summer wood not in clusters.
      a' Color of radial section not yellow..................(Nos. 39–44) ASH.
      b.' Color of radial section light yellow; by which, together with its hardness and weight, this species is easily recognized..(No. 103) OSAGE ORANGE.
    b. Wood light and soft; pores in the summer wood in clusters of 10 to 30, (No. 56) CATALPA.
  2. Pith rays very fine, yet distinct; pores in summer wood usually single or in short lines; color of heartwood reddish brown; of sapwood yellowish white; peculiar odor on fresh section ....(No. 111) SASSAFRAS.
  3. Pith rays fine, but distinct.
    a. Very heavy and hard; heartwood yellowish brown. (No. 77) BLACK LOCUST.
    b. Heavy; medium hard to hard.
      a.' Pores in summer wood very minute, usually in small clusters of 3 to 8; heartwood light orange brown .... (No. 83) RED MULBERRY.
      b.' Pores in summer wood small to minute, usually isolated; heartwood cherry red.................................. (No. 61) COFFEE TREE.

ADDITIONAL NOTES FOR DISTINCTIONS IN THE GROUP.

Sassafras and mulberry may be confounded but for the greater weight and hardness and the absence of odor in the mulberry; the radial section of mulberry also shows the pith rays conspicuously.

Honey locust, coffee tree, and black locust are also very similar in appearance. The honey locust stands out by the conspicuousness of the pith rays, especially on radial sections, on account of their height, while the black locust is distinguished by the extremely great weight and hardness, together with its darker brown color.

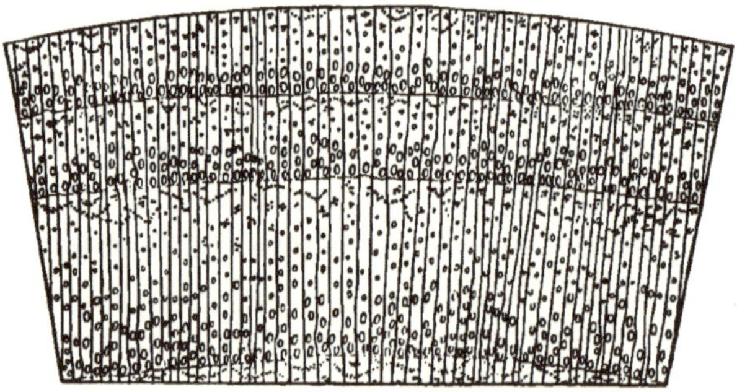

FIG. 40.—Wood of coffee tree.

The ashes, elms, hickories, and oaks may, on casual observation, appear to resemble one another on account of the pronounced zone of porous spring wood. The sharply defined large pith rays of the oak exclude these at once; the wavy lines of pores in the summer wood, appearing as conspicuous finely-feathered hatchings on tangential section, distinguish the elms; while the ashes differ from the hickory by the very conspicuously defined zone of spring-wood pores, which in hickory appear more or less interrupted. The reddish hue of the hickory and the more or less brown hue of the ash may also aid in ready recognition. The smooth, radial surface of split hickory will readily separate it from the rest.

4. Pith rays fine but very conspicuous, even without magnifier. **Color of heart-wood red; of sapwood pale lemon** .... (No. 78) HONEY LOCUST.

B. Pores of summer wood minute or small, in concentric wavy and sometimes branching lines, appearing as finely-feathered hatchings on tangential section.

    1. Pith rays fine, but very distinct; color greenish white. Heartwood absent or imperfectly developed................................(No. 70) HACKBERRY.

    2. Pith rays indistinct; color of heartwood reddish brown ; sapwood grayish to reddish white ---------------:--------------------------(Nos. 62–66) ELMS.

C. Pores of summer wood arranged in radial branching lines (when very crowded radial arrangement somewhat obscured).

    1. Pith rays very minute, hardly visible.................(Nos. 58–60) CHESTNUT.

    2. Pith rays very broad and conspicuous........................(Nos. 84–102) OAK.

D. Pores of summer wood mostly but little smaller than those of the spring wood, isolated and scattered; very heavy and hard woods. The pores of the spring wood sometimes form but an imperfect zone. (Some diffuse-porous woods of groups A and B may seem to belong here.)

    1. Fine concentric lines (not of pores) as distinct, or nearly so, as the very fine pith rays; outer summer wood with a tinge of red; heartwood light reddish brown ....(Nos. 71–75) HICKORY.

    2. Fine concentric lines, much finer than the pith rays; no reddish tinge in summer wood; sapwood white; heartwood blackish,

                                                     (No. 105) PERSIMMON.

ADDITIONAL NOTES FOR DISTINCTIONS IN THE GROUP.

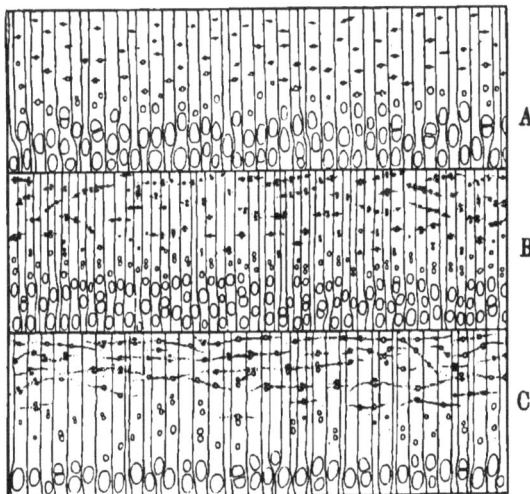

FIG. 41.—*A*, black ash; *B*, white ash; *C*, green ash.

The different species of ash may be identified as follows:

    1. Pores in the summer wood more or less united into lines.

      a. The lines short and broken, occurring mostly near the limit of the ring .... (No 39) WHITE ASH.

      b. The lines quite long and conspicuous in most parts of the summer wood ....(No. 43) GREEN ASH.

    2. Pores in the summer wood not united into lines, or rarely so.

      a. Heartwood reddish brown and very firm .... (No. 40) RED ASH.

      b. Heartwood grayish brown, and much more porous.. (No. 41) BLACK ASH.

In the oaks, two groups can be readily distinguished by the manner in which the pores are distributed in the summer wood. In the white oaks the pores are very fine and numerous and crowded in the outer part of the summer wood, while in the black or red oaks the pores are larger, few in number, and mostly isolated. The live oaks, as far as structure is concerned, belong to the black oaks, but are much less porous, and are exceedingly heavy and hard. ·

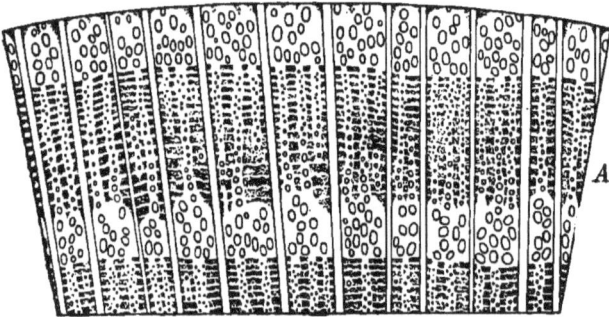

A

FIG. 42.— Wood of red oak. (For white oak see fig. 38.)

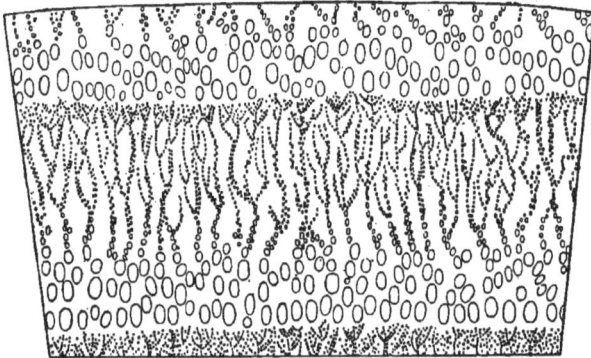

FIG. 43.— Wood of chestnut.

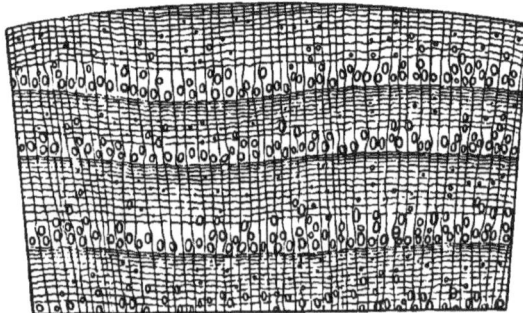

FIG. 44.—Wood of hickory.

ANALYTICAL KEY. 69

III.—DIFFUSE-POROUS WOODS.

[A few indistinctly ring-porous woods of Group II, D, and cedar elm may seem to belong here.]

A. Pores varying in size from large to minute; largest in spring wood, thereby giving sometimes the appearance of a ring-porous arrangement.

  1. Heavy and hard; color of heartwood (especially on longitudinal section) chocolate brown .... (No. 116) BLACK WALNUT.

  2. Light and soft; color of heartwood light reddish brown.(No. 55) BUTTERNUT.

B. Pores all minute and indistinct; most numerous in spring wood, giving rise to a lighter colored zone or line (especially on longitudinal section), thereby appearing sometimes ring-porous; wood hard, heartwood vinous reddish; pith rays very fine, but very distinct. (See also the sometimes indistinct ring-porous cedar elm, and occasionally winged elm, which are readily distinguished by the concentric wavy lines of pores in the summer wood) ....(No. 57) CHERRY.

C. Pores minute or indistinct, neither conspicuously larger nor more numerous in the spring wood and evenly distributed.

  1. Broad pith rays present.

    a. All or most pith rays broad, numerous, and crowded, especially on tangential sections, medium heavy and hard, difficult to split,

        (Nos. 112 and 113) SYCAMORE.

    b. Only part of the pith rays broad.

      a'. Broad pith rays well defined, quite numerous; wood reddish-white to reddish..(No. 47) BEECH.

      b'. Broad pith rays not sharply defined, made up of many small rays, not numerous. Stem furrowed, and therefore the periphery of section, and with it the annual rings sinuous, bending in and out, and the large pith rays generally limited to the furrows or concave portions. Wood white, not reddish............... (No. 52) BLUE BEECH.

  2. No broad pith rays present.

    a. Pith rays small to very small, but quite distinct.

      a'. Wood hard.

        a''. Color reddish white, with dark reddish tinge in outer summer wood.......................... ....(Nos. 79–82) MAPLE.

        b''. Color white, without reddish tinge ....(No. 76) HOLLY.

      b'. Wood soft to very soft.

        a''. Pores crowded, occupying nearly all the space between pith rays.

          a'''. Color yellowish white, often with greenish tinge in heartwood .... (No. 115) TULIP POPLAR,

              (No. 116) CUCUMBER TREE,

          b'''. Color of sapwood grayish, of heartwood light to dark reddish brown............................. (No. 69) SWEET GUM.

        b''. Pores not crowded, occupying not over one-third the space between pith rays; heartwood brownish white to very light brown,

          (Nos. 45 and 46) BASSWOOD.

    b. Pith rays scarcely distinct, yet if viewed with ordinary magnifier, plainly visible.

      a'. Pores indistinct to the naked eye.

        a''. Color uniform pale yellow; pith rays not conspicuous even on the radial section ....(Nos. 53 and 54) BUCKEYE.

        b''. Sapwood yellowish gray, heartwood grayish brown; pith rays conspicuous on the radial section ....(Nos. 67–68) SOUR GUM.

      b'. Pores scarcely distinct,but mostly visible as grayish specks on the cross section; sapwood whitish, heartwood reddish....(Nos. 48–51) BIRCH.

D. Pith rays not visible or else indistinct, even if viewed with magnifier.

  1. Wood very soft, white, or in shades of brown, usually with a silky luster,

    (Nos. 105–110) COTTONWOOD (POPLAR.)

ADDITIONAL NOTES FOR DISTINCTIONS IN THE GROUP.

Cherry and birch are sometimes confounded, the high pith rays on the cherry on radial sections readily distinguishes it; distinct pores on birch and spring-wood zone in cherry as well as the darker vinous-brown color of the latter will prove helpful.

Two groups of birches can be readily distinguished, though specific distinction is not always possible.

    1. Pith rays fairly distinct, the pores rather few and not more abundant in the spring wood; wood heavy, usually darker,

(No. 48) CHERRY BIRCH and (No. 49) YELLOW BIRCH.

    2. Pith rays barely distinct, pores more numerous and commonly forming a more porous spring-wood zone; wood of medium weight,

- (No. 51) CANOE OR PAPER BIRCH.

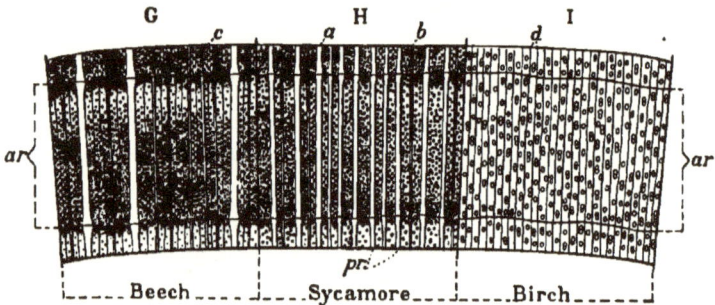

FIG. 45.—Wood of beech, sycamore, and birch.

The species of maple may be distinguished as follows:

    1. Most of the pith rays broader than the pores and very conspicuous,

(No. 79) SUGAR MAPLE.

    2. Pith rays not or rarely broader than the pores, fine but conspicuous.

        a. Wood heavy and hard, usually of darker reddish color and commonly spotted on cross section...................... (No. 80) RED MAPLE.

        b. Wood of medium weight and hardness, usually light colored,

(No. 82) SILVER MAPLE.

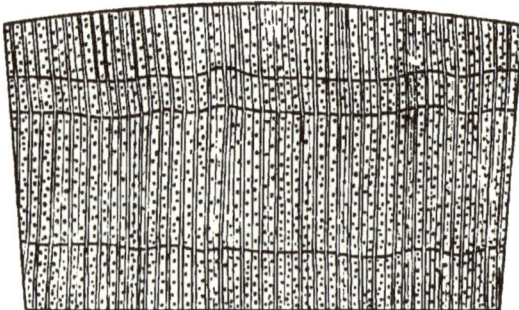

FIG. 46.—Wood of maple.

Red maple is not always safely distinguished from soft maple. In box elder the pores are finer and more numerous than in soft maple.

The various species of elm may be distinguished as follows:

1. Pores of spring wood form a broad band of several rows; easy splitting, dark brown heart .... (No. 64) RED ELM.
2. Pores of spring wood usually in a single row, or nearly so.
    a. Pores of spring wood large, conspicuously so ...... (No. 62) WHITE ELM.
    b. Pores of spring wood small to minute.
        a'. Lines of pores in summer wood fine, not as wide as the intermediate spaces, giving rise to very compact grain .... (No. 63) ROCK ELM.
        b'. Lines of pores broad, commonly as wide as the intermediate spaces, (No. 66) WINGED ELM.
    c. Pores in spring wood indistinct, and therefore hardly a ring-porous wood .... (No. 65) CEDAR ELM.

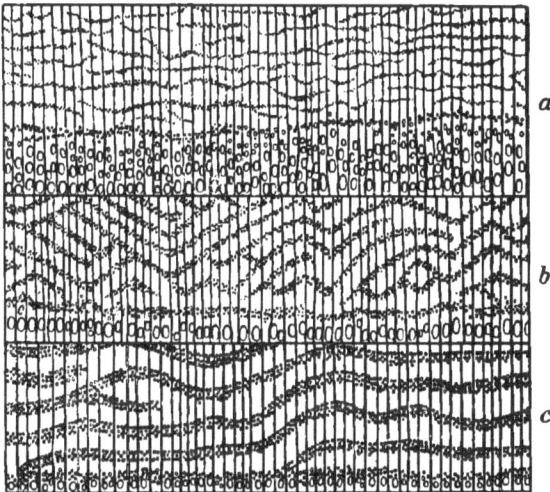

FIG. 47.—Wood of elm. *a*, red elm; *b*, white elm; *c*, winged elm.

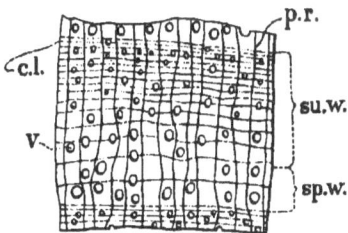

FIG. 48.—Walnut. *p. r.*, pith rays; *c. l.*, concentric lines; *v*, vessels or pores; *su. w.*, summer wood; *sp. w.*, spring wood.

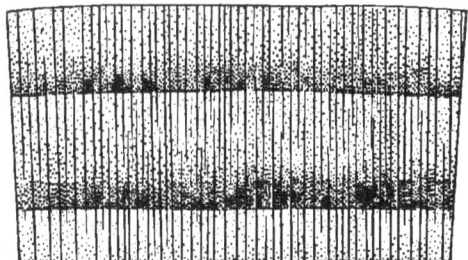

FIG. 49.—Wood of cherry.

LIST OF THE MORE IMPORTANT WOODS OF THE UNITED STATES.

[Arranged alphabetically.]

A.—CONIFEROUS WOODS.

Woods of simple and uniform structure, generally light, soft but stiff; abundant in suitable dimensions and forming by far the greatest part of all the lumber used. CEDAR.—Light, soft, stiff, not strong, of fine texture; sap and heartwood distinct, the former lighter, the latter a dull, grayish brown, or red. The wood seasons rapidly, shrinks and checks but little, and is very durable. Used like soft pine, but owing to its great durability preferred for shingles, etc. Small sizes used for posts, ties, etc.[1] Cedars usually occur scattered, but they form, in certain localities, forests of considerable extent.

    a. **White cedars.**—Heartwood a light grayish brown.

1. WHITE CEDAR (*Thuya occidentalis*) (Arborvitæ): Scattered along streams and lakes, frequently covering extensive swamps; rarely large enough for lumber, but commonly used for posts, ties, etc. Maine to Minnesota and northward.
2. CANOE CEDAR (*Thuya gigantea*)(red cedar of the West): In Oregon and Washington a very large tree, covering extensive swamps; in the mountains much smaller, skirting the water courses; an important lumber tree. Washington to northern California and eastward to Montana.
3. WHITE CEDAR(*Chamæcyparis thyoides*): Medium-sized tree, wood very light and soft. Along the coast from Maine to Mississippi.
4. WHITE CEDAR (*Chamæcyparis lawsoniana*) (Port Orford cedar, Oregon cedar, Lawson's cypress, ginger pine): A very large tree, extensively cut for lumber; heavier and stronger than the preceding. Along the coast line of Oregon.
5. WHITE CEDAR (*Libocedrus decurrens*) (incense cedar): A large tree, abundantly scattered among pine and fir; wood fine grained. Cascades and Sierra Nevada of Oregon and California.

    b. **Red cedars.**—Heartwood red.

6. RED CEDAR (*Juniperus virginiana*) (Savin juniper): Similar to white cedar, but of somewhat finer texture. Used in cabinetwork in cooperage, for veneers, and especially for lead pencils, for which purpose alone several million feet are cut each year. A small to medium sized tree scattered through the forests, or, in the West, sparsely covering extensive areas (cedar brakes). The red cedar is the most widely distributed conifer of the United States, occurring from the Atlantic to the Pacific and from Florida to Minnesota, but attains a suitable size for lumber only in the Southern, and more especially the Gulf, States.
7. REDWOOD (*Sequoia sempervirens*): Wood in its quality and uses like white cedar; the narrow sapwood whitish; the heartwood light red, soon turning to brownish red when exposed. A very large tree, limited to the coast ranges of California, and forming considerable forests, which are rapidly being converted into lumber.

CYPRESS.

8. CYPRESS (*Taxodium distichum*) (bald cypress; black, white, and red cypress): Wood in appearance, quality, and uses similar to white cedar. "Black

---

[1] Since almost all kinds of woods are used for fuel and charcoal, and in the construction of fences, sheds, barns, etc., the enumeration of these uses has been omitted in this list.

cypress" and "white cypress" are heavy and light forms of the same species. The cypress is a large deciduous tree, occupying much of the swamp and overflow land along the coast and rivers of the Southern States.

FIR.—This name is frequently applied to wood and to trees which are not fir; most commonly to spruce, but also, especially in English markets, to pine. It resembles spruce, but is easily distinguished from it, as well as from pine and larch, by the absence of resin ducts. Quality, uses, and habits similar to spruce.

9. BALSAM fir (*Abies balsamea*): A medium-sized tree scattered throughout the northern pineries; cut, in lumber operations whenever of sufficient size, and sold with pine or spruce. Minnesota to Maine and northward.

10. WHITE FIR (*Abies grandis* and *Abies concolor*): Medium to very large sized tree, forming an important part of most of the Western mountain forests, and furnishing much of the lumber of the respective regions. The former occurs from Vancouver to central California and eastward to Montana; the latter from Oregon to Arizona and eastward to Colorado and New Mexico.

11. WHITE FIR (*Abies amabilis*): Good-sized tree, often forming extensive mountain forests. Cascade Mountains of Washington and Oregon.

12. RED FIR (*Abies nobilis*) (not to be confounded with Douglas fir; see No. 37): Large to very large tree, forming with *A. amabilis* extensive forests on the slope of the mountains between 3,000 and 4,000 feet elevation. Cascade Mountains of Oregon.

13. RED FIR (*Abies magnifica*): Very large tree, forming forests about the base of Mount Shasta. Sierra Nevada of California, from Mount Shasta southward.

HEMLOCK.—Light to medium weight, soft, stiff but brittle, commonly crossgrained, rough and splintery; sapwood and heartwood not well defined; the wood of a light, reddish-gray color, free from resin ducts, moderately durable, shrinks and warps considerably, wears rough, retains nails firmly. Used principally for dimension stuff and timbers. Hemlocks are medium to large sized trees, commonly scattered among broad-leaved trees and conifers, but often forming forests of almost pure growth.

14. HEMLOCK (*Tsuga canadensis*): Medium-sized tree, furnishes almost all the hemlock of the Eastern market. Maine to Wisconsin; also following the Alleghanies southward to Georgia and Alabama.

15. HEMLOCK (*Tsuga mertensiana*): Large-sized tree, wood claimed to be heavier and harder than the Eastern form and of superior quality. Washington to California and eastward to Montana.

LARCH OR TAMARACK.—Wood like the best of hard pine, both in appearance, quality, and uses, and owing to its great durability, somewhat preferred in shipbuilding, for telegraph poles, and railroad ties. In its structure it resembles spruce. The larches are deciduous trees, occasionally covering considerable areas, but usually scattered among other conifers.

16. TAMARACK (*Larix americana*) (Hackmatack): Medium-sized tree, often covering swamps, in which case it is smaller and of poor quality. Maine to Minnesota, and southward to Pennsylvania.

17. TAMARACK (*L. occidentalis*): Large-sized trees, scattered, locally abundant. Washington and Oregon to Montana.

PINE.—Very variable, very light and soft in "soft" pine, such as white pine; of medium weight to heavy and quite hard in "hard" pine, of which longleaf or Georgia pine is the extreme form. Usually it is stiff, quite strong, of even texture, and more or less resinous. The sapwood is yellowish white; the heartwood, orange brown. Pine shrinks moderately, seasons rapidly and without much injury; it works easily; is never too hard to nail (unlike oak or hickory); it is mostly quite durable, and if well seasoned is not subject to the attacks of boring insects. The heavier the wood, the darker, stronger, and harder it is, and the more it shrinks and checks. Pine is used more extensively than any other kind of wood. It is the principal wood in common carpentry, as well as in all heavy construction, bridges, trestles, etc. It is also used in almost every other wood industry, for spars, masts, planks, and timbers in shipbuilding, in car and wagon construction, in cooperage, for crates and boxes, in furniture work, for toys and patterns, railway ties, water pipes, excelsior, etc. Pines are usually large trees with few branches, the straight, cylindrical, useful stem forming by far the greatest part of the tree; they occur gregariously, forming vast forests, a fact which greatly facilitates their exploitation. Of the many special terms applied to pine as lumber, denoting sometimes differences in quality, the following deserve attention:

"White pine," "pumpkin pine," "soft pine," in the Eastern markets refer to the wood of the white pine (*Pinus strobus*), and on the Pacific Coast to that of the sugar pine (*Pinus lambertiana*).

"Yellow pine" is applied in the trade to all the Southern lumber pines; in the Northeast it is also applied to the pitch pine (*P. rigida*); in the West it refers mostly to bull pine (*P. ponderosa*).

"Yellow longleaf pine," "Georgia pine," chiefly used in advertisement, refers to longleaf pine (*P. palustris*).

"Hard pine" is a common term in carpentry, and applies to everything except white pine.

"Pitch pine" includes all Southern pines and also the true pitch pine (*P. rigida*), but is mostly applied, especially in foreign markets, to the wood of the longleaf pine (*P. palustris*).

For the great variety of confusing local names applied to the Southern pines in their homes, part of which have been adopted in the markets of the Atlantic seaboard, see report of Chief of Division of Forestry for 1891, page 212, etc., and also the list below:

    a. Soft pines.

18. WHITE PINE (*Pinus strobus*): Large to very large sized tree; for the last fifty years the most important timber tree of the Union, furnishing the best quality of soft pine. Minnesota, Wisconsin, Michigan, New England, along the Alleghanies to Georgia.

19. SUGAR PINE (*Pinus lambertiana*): A very large tree, together with *Abies concolor*, forming extensive forests; important lumber tree. Oregon and California.

20. WHITE PINE (*Pinus monticola*): A large tree, at home in Montana, Idaho, and the Pacific States; most common and locally used in northern Idaho.

21. WHITE PINE (*Pinus flexilis*): A small tree, forming mountain forests of considerable extent and locally used; Eastern Rocky Mountain slopes; Montana to New Mexico.

    b. Hard pines.

22. LONGLEAF PINE (*Pinus palustris*) (Georgia pine, yellow pine, long straw pine, etc.): Large tree; forms extensive forests and furnishes the hardest and strongest pine lumber in the market. Coast region from North Carolina to Texas.

23. BULL PINE (*Pinus ponderosa*) (yellow pine): Medium to very large sized tree, forming extensive forests in Pacific and Rocky Mountain regions; furnishes most of the hard pine of the West; sapwood wide; wood very variable.

24. LOBLOLLY PINE (*Pinus tæda*) (slash pine, old field pine, rosemary pine, sap pine, short straw pine, etc.): Large-sized tree, forms extensive forests; wider-ringed, coarser, lighter, softer, with more sapwood than the longleaf pine, but the two often confounded. This is the common lumber pine from Virginia to South Carolina, and is found extensively in Arkansas and Texas. Southern States; Virginia to Texas and Arkansas.

25. NORWAY PINE (*Pinus resinosa*): Large-sized tree, never forming forests, usually scattered or in small groves, together with white pine; largely sapwood and hence not durable. Minnesota to Michigan; also in New England to Pennsylvania.

26. SHORTLEAF PINE (*Pinus echinata*) (slash pine, Carolina pine, yellow pine, old field pine, etc.): Resembles loblolly pine; often approaches in its wood the Norway pine. The common lumber pine of Missouri and Arkansas. North Carolina to Texas and Missouri.

27. CUBAN PINE (*Pinus cubensis*) (slash pine, swamp pine, bastard pine, meadow pine): Resembles longleaf pine, but commonly has wider sapwood and coarser grain; does not enter the markets to any great extent. Along the coast from South Carolina to Louisiana.

28. BULL PINE (*Pinus jeffreyi*) (black pine): Large-sized tree, wood resembling bull pine (*P. ponderosa*); used locally in California, replacing *P. ponderosa* at high altitudes.

The following are small to medium sized pines, not commonly offered as lumber in the market; used locally for timber, ties, etc.:

29. BLACK PINE (*Pinus murrayana*) (lodge-pole pine, tamarack): Rocky Mountains and Pacific regions.

30. PITCH PINE (*Pinus rigida*): Along the coast from New York to Georgia and along the mountains to Kentucky.

31. JERSEY PINE (*Pinus inops*) (scrub pine): As before.

32. GRAY PINE (*Pinus banksiana*) (scrub pine): Maine, Vermont, and Michigan to Minnesota.

REDWOOD. (*See* CEDAR.)

SPRUCE.—Resembles soft pine, is light, very soft, stiff, moderately strong, less resinous than pine; has no distinct heartwood, and is of whitish color. Used like soft pine, but also employed as resonance wood and preferred for paper pulp. Spruces, like pines, form extensive forests; they are more frugal, thrive on thinner soils, and bear more shade, but usually require a more humid climate. "Black" and "white spruce," as applied by lumbermen, usually refer to narrow and wide ringed forms of the black spruce (*Picea nigra*).

33. BLACK SPRUCE (*Picea nigra*): Medium-sized tree, forms extensive forests in northeastern United States and in British America; occurs scattered or in groves, especially in low lands throughout the Northern pineries. Important lumber tree in Eastern United States. Maine to Minnesota, British America, and on the Alleghanies to North Carolina.

34. WHITE SPRUCE (*Picea alba*): Generally associated with the preceding; most abundant along streams and lakes, grows largest in Montana and forms the most important tree of the subarctic forest of British America. Northern United States, from Maine to Minnesota, also from Montana to Pacific, British America.

35. WHITE SPRUCE (*Picea engelmanni*): Medium to large sized tree, forming extensive forests at elevations from 5,000 to 10,000 feet above sea level; resembles the preceding, but occupies a different station. A very important timber tree in the central and southern parts of the Rocky Mountains. Rocky Mountains from Mexico to Montana.

36. TIDE-LAND SPRUCE (*Picea sitchensis*): A large-sized tree, forming an extensive coast-belt forest. Along the seacoast from Alaska to Central California.

BASTARD SPRUCE.—Spruce or fir in name but resembling hard pine or larch in the appearance, quality, and uses of its wood.

37. DOUGLAS SPRUCE (*Pseudotsuga douglasii*) (yellow fir, red fir, Oregon pine): One of the most important trees of the Western United States; grows very large in the Pacific States, to fair size in all parts of the mountains, in Colorado up to about 10,000 feet above sea level; forms extensive forests, often of pure growth. Wood very variable, usually coarsegrained and heavy, with very pronounced summer wood, hard and strong ("red" fir), but often fine-grained and light ("yellow" fir). It replaces hard pine and is especially suited to heavy construction. From the plains to the Pacific Ocean; from Mexico to British America.

TAMARACK. (*See* LARCH.)

YEW.—Wood heavy, hard, extremely stiff and strong, of fine texture with a pale yellow sapwood, and an orange red heart; seasons well and is quite durable. Yew is extensively used for archery, bows, turner's ware, etc. The yews form no forests, but occur scattered with other conifers.

38. YEW (*Taxus brevifolia*): A small to medium sized tree of the Pacific region.

B.—BROAD-LEAVED WOODS (HARDWOODS).

Woods of complex and very variable structure and therefore differing widely in quality, behavior, and consequently in applicability to the arts.

ASH.—Wood heavy, hard, strong, stiff, quite tough, not durable in contact with soil, straight grained, rough on the split surface and coarse in texture. The wood shrinks moderately, seasons with little injury, stands well and takes a good polish. In carpentry ash is used for finishing lumber, stairways, panels, etc.; it is used in shipbuilding, in the construction of cars, wagons, carriages, etc., in the manufacture of farm implements, machinery, and especially of furniture of all kinds, and also for harness work; for barrels, baskets, oars, tool handles, hoops, clothespins, and toys. The trees of the several species of ash are rapid growers, of small to medium height with stout trunks; they form no forests, but occur scattered in almost all our broad-leaved forests.

39. WHITE ASH (*Fraxinus americana*): Medium, sometimes large sized tree. Basin of the Ohio, but found from Maine to Minnesota and Texas.

40. RED ASH (*Fraxinus pubescens*): Small-sized tree. North Atlantic States, but extends to the Mississippi.

41. BLACK ASH (*Fraxinus sambucifolia*) (hoop ash, ground ash): Medium-sized tree, very common. Maine to Minnesota, and southward to Virginia and Arkansas.

42. BLUE ASH (*Fraxinus quadrangulata*): Small to medium sized. Indiana and Illinois; occurs from Michigan to Minnesota and southward to Alabama.

43. GREEN ASH (*Fraxinus viridis*): Small-sized tree. New York to the Rocky Mountains, and southward to Florida and Arizona.

44. OREGON ASH (*Fraxinus oregana*): Medium-sized tree. Western Washington to California.

ASPEN. (*See* POPLAR.)

BASSWOOD.

45. BASSWOOD (*Tilia americana*) (lime tree, American linden, lin, bee tree): Wood light, soft, stiff but not strong, of fine texture, and white to light brown color. The wood shrinks considerably in drying, works and stands well; it is used in carpentry, in the manufacture of furniture and woodenware, both turned and carved, in cooperage, for toys, also for paneling of car and carriage bodies. Medium to large sized tree, common in all Northern broad-leaved forests; found throughout the Eastern United States.

46. WHITE BASSWOOD (*Tilia heterophylia*): A small-sized tree most abundant in the Alleghany region.

BEECH.

47. BEECH (*Fagus ferruginea*): Wood heavy, hard, stiff, strong, of rather coarse texture, white to light brown, not durable in the ground, and subject to the inroads of boring insects; it shrinks and checks considerably in drying, works and stands well and takes a good polish. Used for furniture, in turnery, for handles, lasts, etc. Abroad it is very extensively employed by the carpenter, millwright, and wagon maker, in turnery as well as wood carving. The beech is a medium-sized tree, common, sometimes forming forest; most abundant in the Ohio and Mississippi basin, but found from Maine to Wisconsin and southward to Florida.

BIRCH.—Wood heavy, hard, strong, of fine texture; sapwood whitish, heartwood in shades of brown with red and yellow; very handsome, with satiny luster, equaling cherry. The wood shrinks considerably in drying, works and stands well and takes a good polish, but is not durable, if exposed. Birch is used for finishing lumber in building, in the manufacture of furniture, in wood turnery for spools, boxes, wooden shoes, etc., for shoe lasts and pegs, for wagon hubs, or yokes, etc., also in wood carving. The birches are medium-sized trees, form extensive forests northward and occur scattered in all broad-leaved forests of the Eastern United States.

48. CHERRY BIRCH (*Betula lenta*) (black birch, sweet birch, mahogany birch): Medium-sized tree; very common. Maine to Michigan and to Tennessee.

49. YELLOW BIRCH (*Betula lutea*) (gray birch): Medium-sized tree; common. Maine to Minnesota and southward to Tennessee.

50. RED BIRCH (*Betula nigra*) (river birch): Small to medium sized tree; very common; lighter and less valuable than the preceding. New England to Texas and Missouri.

51. CANOE BIRCH (*Betula papyrifera*) (white birch, paper birch): Generally a small tree; common, forming forests; wood of good quality but lighter. All along the northern boundary of United States and northward, from the Atlantic to the Pacific.

BLACK WALNUT. (*See* WALNUT.)

BLUE BEECH.

52. BLUE BEECH (*Carpinus caroliniana*) (hornbeam, water beech, ironwood): Wood very heavy, hard, strong, very stiff, of rather fine texture and white color; not durable in the ground; shrinks and checks greatly, but works and stands well. Used chiefly in turnery for tool handles, etc. Abroad, much used by mill and wheel wrights. A small tree, largest in the Southwest, but found in nearly all parts of the Eastern United States.

BOIS D'ARC. (*See* OSAGE ORANGE.)

BUCKEYE—HORSE CHESTNUT.—Wood light, soft, not strong, often quite tough, of fine and uniform texture and creamy white color. It shrinks considerably, but works and stands well. Used for wooden ware, artificial limbs, paper pulp, and locally also for building lumber. Small-sized trees, scattered.

53. OHIO BUCKEYE (*Æsculus glabra*) (fetid buckeye): Alleghanies, Pennsylvania to Indian Territory.

54. SWEET BUCKEYE (*Æsculus flava*): Alleghanies, Pennsylvania to Texas.

BUTTERNUT.

55. BUTTERNUT (*Juglans cinerea*) (white walnut): Wood very similar to black walnut, but light, quite soft, not strong and of light brown color. Used chiefly for finishing lumber, cabinetwork, and cooperage. Medium-sized tree, largest and most common in the Ohio basin; Maine to Minnesota and southward to Georgia and Alabama.

CATALPA.

56. CATALPA (*Catalpa speciosa*): Wood light, soft, not strong, brittle, durable, of coarse texture and brown color; used for ties and posts, but well suited for a great variety of uses. Medium-sized tree; lower basin of the Ohio River, locally common. Extensively planted, and therefore promising to become of some importance.

CHERRY.

57. CHERRY (*Prunus serotina*): Wood heavy, hard, strong, of fine texture; sapwood yellowish white, heartwood reddish to brown. The wood shrinks considerably in drying, works and stands well, takes a good polish, and is much esteemed for its beauty. Cherry is chiefly used as a decorative finishing lumber for buildings, cars, and boats, also for furniture and in turnery. It is becoming too costly for many purposes for which it is naturally well suited. The lumber-furnishing cherry of this country, the wild black cherry (*Prunus serotina*), is a small to medium sized tree, scattered through many of the broad-leaved woods of the western slope of the Alleghanies, but found from Michigan to Florida and west to Texas. Other species of this genus as well

as the hawthorns (*Cratægus*) and wild apple (*Pyrus*) are not commonly offered in the market. Their wood is of the same character as cherry, often even finer, but in small dimensions.

CHESTNUT.

58. CHESTNUT (*Castanea vulgaris* var. *americana*): Wood light, moderately soft, stiff, not strong, of coarse texture; the sapwood light, the heartwood darker brown. It shrinks and checks considerably in drying, works easily, stands well, and is very durable. Used in cabinetwork, cooperage, for railway ties, telegraph poles, and locally in heavy construction. Medium-sized tree, very common in the Alleghanies, occurs from Maine to Michigan and southward to Alabama.

59. CHINQUAPIN (*Castanea pumila*): A small-sized tree, with wood slightly heavier but otherwise similar to the preceding; most common in Arkansas, but with nearly the same range as the chestnut.

60. CHINQUAPIN (*Castanopsis chrysophylla*): A medium-sized tree of the western ranges of California and Oregon.

COFFEE TREE.

61. COFFEE TREE (*Gymnocladus canadensis*) (coffee nut): Wood heavy, hard, strong, very stiff, of coarse texture, durable; the sapwood yellow, the heartwood reddish brown; shrinks and checks considerably in drying; works and stands well and takes a good polish. It is used to a limited extent in cabinetwork. A medium to large sized tree; not common. Pennsylvania to Minnesota and Arkansas.

COTTONWOOD. (*See* POPLAR.)
CUCUMBER TREE. (*See* TULIP.)

ELM.—Wood heavy, hard, strong, very tough; moderately durable in contact with the soil; commonly crossgrained, difficult to split and shape, warps, and checks considerably in drying, but stands well if properly handled. The broad sapwood whitish, heart brown, both with shades of gray and red; on split surface rough; texture coarse to fine; capable of high polish. Elm is used in the construction of cars, wagons, etc., in boat and ship building, for agricultural implements and machinery; in rough cooperage, saddlery and harness work, but particularly in the manufacture of all kinds of furniture, where the beautiful figures, especially those of the tangential or bastard section, are just beginning to be duly appreciated. The elms are medium to large sized trees, of fairly rapid growth, with stout trunk, form no forests of pure growth, but are found scattered in all the broad-leaved woods of our country, sometimes forming a considerable portion of the arborescent growth.

62. WHITE ELM (*Ulmus americana*) (American elm, water elm): Medium to large sized tree, common. Maine to Minnesota, southward to Florida and Texas.

63. ROCK ELM (*Ulmus racemosa*) (cork elm, hickory elm, white elm, cliff elm): Medium to large sized tree. Michigan, Ohio, from Vermont to Iowa, southward to Kentucky.

64. RED ELM (*Ulmus fulva*) (slippery elm, moose elm): Small-sized tree, found chiefly along water courses. New York to Minnesota, and southward to Florida and Texas.

65. CEDAR ELM (*Ulmus crassifolia*): Small-sized tree, quite common. Arkansas and Texas.

66. WINGED ELM (*Ulmus alata*) (Wahoo): Small-sized tree, locally quite common. Arkansas, Missouri, and eastern Virginia.

GUM.—This general term refers to two kinds of wood usually distinguished as sweet or red gum, and sour, black, or tupelo gum, the former being a relative of the witch-hazel, the latter belonging to the dogwood family.

67. TUPELO (*Nyssa sylvatica*) (sour gum, black gum): Maine to Michigan, and southward to Florida and Texas. Wood heavy, hard, strong, tough, of fine texture,

frequently crossgrained, of yellowish or grayish white color, hard to split and work, troublesome in seasoning, warps and checks considerably, and is not durable if exposed; used for wagon hubs, wooden ware, handles, wooden shoes, etc. Medium to large sized trees, with straight, clear trunks; locally quite abundant, but never forming forests of pure growth.

68. TUPELO GUM (*Nyssa uniflora*) (cotton gum): Lower Mississippi basin, northward to Illinois and eastward to Virginia, otherwise like preceding species.

69. SWEET GUM (*Liquidambar styraciflua*) (red gum, liquidambar, bilsted): Wood rather heavy, rather soft, quite stiff and strong, tough, commonly crossgrained, of fine texture; the broad sapwood whitish, the heartwood reddish brown; the wood shrinks and warps considerably, but does not check badly, stands well when fully seasoned, and takes good polish. Sweet gum is used in carpentry, in the manufacture of furniture, for cut veneer, for wooden plates, plaques, baskets, etc., also for wagon hubs, hat blocks, etc. A large-sized tree, very abundant, often the principal tree in the swampy parts of the bottoms of the Lower Mississippi Valley; occurs from New York to Texas and from Indiana to Florida.

HACKBERRY.

70. HACKBERRY (*Celtis occidentalis*) (sugar berry): The handsome wood heavy, hard, strong, quite tough, of moderately fine texture, and greenish or yellowish white color; shrinks moderately, wor's well, and takes a good polish. So far but little used in the manufacture of furniture. Medium to large sized tree, locally quite common, largest in the Lower Mississippi Valley; occurs in nearly all parts of the Eastern United States.

HICKORY.—Wood very heavy, hard, and strong, proverbially tough, of rather coarse texture, smooth and of straight grain. The broad sapwood white, the heart reddish nut brown. It dries slowly, shrinks and checks considerably; is not durable in the ground, or if exposed, and, especially the sapwood, is always subject to the inroads of boring insects. Hickory excels as carriage and wagon stock, but is also extensively used in the manufacture of implements and machinery, for tool handles, timber pins, for harness work, and cooperage. The hickories are tall trees with slender stems, never form forests, occasionally small groves, but usually occur scattered among other broad-leaved trees in suitable localities. The following species all contribute more or less to the hickory of the markets:

71. SHAGBARK HICKORY (*Hicoria ovata*) (shellbark hickory): A medium to large sized tree, quite common; the favorite among hickories; best developed in the Ohio and Mississippi basins; from Lake Ontario to Texas, Minnesota to Florida.

72. MOCKERNUT HICKORY (*Hicoria alba*) (black hickory, bull and black nut, big bud, and white-heart hickory): A medium to large sized tree, with the same range as the foregoing; common, especially in the South.

73. PIGNUT HICKORY (*Hicoria glabra*) (brown hickory, black hickory, switch-bud hickory): Medium to large sized tree, abundant; all Eastern United States.

74. BITTER NUT HICKORY (*Hicoria minima*) (swamp hickory): A medium-sized tree, favoring wet localities, with the same range as the preceding.

75. PECAN (*Hicoria pecan*) (Illinois nut): A large tree, very common in the fertile bottoms of the Western streams. Indiana to Nebraska and southward to Lousiana and Texas.

HOLLY.

76. HOLLY (*Ilex opaca*): Wood of medium weight, hard, strong, tough, of fine texture and white color; works and stands well, used for cabinetwork and turnery. A small tree, most abundant in the Lower Mississippi Valley and Gulf States, but occurring eastward to Massachusetts and north to Indiana.

HORSE-CHESTNUT. (*See* BUCKEYE.)

IRONWOOD. (*See* BLUE BEECH.)

LOCUST.—This name applies to both of the following:

77. BLACK LOCUST (*Robinia pseudacacia*) (black locust, yellow locust): Wood very heavy, hard, strong, and tough, of coarse texture, very durable in contact with the soil, shrinks considerably and suffers in seasoning; the very narrow sapwood yellowish, the heartwood brown, with shades of red and green. Used for wagon hubs, tree nails or pins, but especially for ties, posts, etc. Abroad it is much used for furniture and farm implements and also in turnery. Small to medium sized tree, at home in the Alleghanies, extensively planted, especially in the West.

78. HONEY LOCUST (*Gleditschia triacanthos*) (black locust, sweet locust, three-thorned acacia): Wood heavy, hard, strong, tough, of coarse texture, susceptible of a good polish, the narrow sapwood yellow, the heartwood brownish red. So far, but little appreciated except for fencing and fuel; used to some extent for wagon hubs and in rough construction. A medium-sized tree, found from Pennsylvania to Nebraska, and southward to Florida and Texas; locally quite abundant.

MAGNOLIA. (*See* TULIP.)

MAPLE.—Wood heavy, hard, strong, stiff, and tough, of fine texture, frequently wavy-grained, this giving rise to "curly" and "blister" figures; not durable in the ground or otherwise exposed. Maple is creamy white, with shades of light brown in the heart; shrinks moderately, seasons, works and stands well, wears smoothly, and takes a fine polish. The wood is used for ceiling, flooring, paneling, stairway, and other finishing lumber in house, ship, and car construction; it is used for the keels of boats and ships, in the manufacture of implements and machinery, but especially for furniture, where entire chamber sets of maple rival those of oak. Maple is also used for shoe lasts and other form blocks, for shoe pegs, for piano actions, school apparatus, for wood type in show bill printing, tool handles, in wood carving, turnery, and scroll work. The maples are medium-sized trees, of fairly rapid growth; sometimes form forests and frequently constitute a large proportion of the arborescent growth.

79. SUGAR MAPLE (*Acer saccharum*) (hard maple, rock maple): Medium to large sized tree, very common, forms considerable forests. Maine to Minnesota, abundant, with birch, in parts of the pineries; southward to northern Florida; most abundant in the region of the Great Lakes.

80. RED MAPLE (*Acer rubrum*) (swamp or water maple): Medium-sized tree. Like the preceding, but scattered along water courses and other moist localities

81. SILVER MAPLE (*Acer saccharinum*) (soft maple, silver maple): Medium-sized, common; wood lighter, softer, inferior to hard maple, and usually offered in small quantities and held separate in the market. Valley of the Ohio, but occurs from Maine to Dakota and southward to Florida.

82. BROAD-LEAFED MAPLE (*Acer macrophyllum*): Medium-sized tree, forms considerable forests, and like the preceding has a lighter, softer, and less valuable wood. Pacific Coast.

MULBERRY.

83. RED MULBERRY (*Morus rubra*): Wood moderately heavy, hard, strong, rather tough, of coarse texture, durable; sapwood whitish, heart yellow to orange brown; shrinks and checks considerably in drying; works and stands well. Used in cooperage and locally in shipbuilding and in the manufacture of farm implements. A small-sized tree, common in the Ohio and Mississippi valleys, but widely distributed in the Eastern United States.

OAK.—Wood very variable, usually very heavy and hard, very strong and tough, porous, and of coarse texture; the sapwood whitish, the heart "oak" brown to reddish brown. It shrinks and checks badly, giving trouble in seasoning, but stands well, is durable, and little subject to attacks of insects. Oak is used for many purposes: in shipbuilding, for heavy construction, in common carpentry, in furniture, car, and wagon work, cooperage, turnery, and even in wood carving; also in the manufacture of all kinds of farm implements, wooden mill machinery, for piles and wharves, railway ties, etc. The oaks are medium to large sized trees, forming the predominant part of a large portion of our broad-leaved forests, so that these are generally "oak forests" though they always contain a considerable proportion of other kinds of trees. Three well-marked kinds, white, red, and live oak, are distinguished and kept separate in the market. Of the two principal kinds white oak is the stronger, tougher, less porous, and more durable. Red oak, is usually of coarser texture, more porous, often brittle, less durable, and even more troublesome in seasoning than white oak. In carpentry and furniture work, red oak brings about the same price at present as white oak. The red oaks everywhere accompany the white oaks, and, like the latter, are usually represented by several species in any given locality. Live oak, once largely employed in shipbuilding, possesses all the good qualities (except that of size) of white oak, even to a greater degree. It is one of the heaviest, hardest, and most durable building timbers of this country; in structure it resembles the red oaks, but is much less porous.

84. WHITE OAK (*Quercus alba*): Medium to large sized tree, common in the Eastern States, Ohio and Mississippi valleys; occurs throughout Eastern United States.

85. BUR OAK (*Quercus macrocarpa*) (mossy-cup oak, over-cup oak): Large-sized tree, locally abundant, common. Bottoms west of Mississippi; range farther west than preceding.

86. SWAMP WHITE OAK (*Quercus bicolor*): Large-sized tree, common. Most abundant in the Lake States, but with range as in white oak.

87. YELLOW OAK (*Quercus prinoides*) (chestnut oak, chinquapin oak): Medium-sized tree. Southern Alleghanies, eastward to Massachusetts.

88. BASKET OAK (*Quercus michauxii*) (cow oak): Large-sized tree, locally abundant; lower Mississippi and eastward to Delaware.

89. OVER-CUP OAK (*Quercus lyrata*) (swamp white oak, swamp post oak): Medium to large sized tree, rather restricted; ranges as in the preceding.

90. POST OAK (*Quercus obtusiloba*) (iron oak): Medium to large sized tree. Arkansas to Texas, eastward to New England and northward to Michigan.

91. WHITE OAK (*Quercus durandii*): Medium to small sized tree. Texas, eastward to Alabama.

92. WHITE OAK (*Quercus garryana*): Medium to large sized tree. Washington to California.

93. WHITE OAK (*Quercus lobata*): Medium to large-sized tree; largest oak on the Pacific Coast; California.

94. RED OAK (*Quercus rubra*) (black oak): Medium to large sized tree; common in all parts of its range. Maine to Minnesota, and southward to the Gulf.

95. BLACK OAK (*Quercus tinctoria*), (yellow oak): Medium to large sized tree; very common in the Southern States, but occurring north as far as Minnesota, and eastward to Maine.

96. SPANISH OAK (*Quercus falcata*), (red oak): Medium sized tree, common in the South Atlantic and Gulf region, but found from Texas to New York, and north to Missouri and Kentucky.

97. SCARLET OAK (*Quercus coccinea*): Medium to large sized tree; best developed in the lower basin of the Ohio, but found from Maine to Missouri, and from Minnesota to Florida.

98. PIN OAK (*Quercus palustris*) (swamp spanish oak, water oak): Medium to large sized tree, common along borders of streams and swamps. Arkansas to Wisconsin, and eastward to the Alleghanies.

99. WILLOW OAK (*Quercus phellos*) (peach oak): Small to medium sized tree. New York to Texas, and northward to Kentucky.

3521—No. 10——6

100. WATER OAK (*Quercus aquatica*) (duck oak, possum oak, punk oak): Medium to large sized tree, of extremely rapid growth. Eastern Gulf States, eastward to Delaware, and northward to Missouri and Kentucky.

101. LIVE OAK (*Quercus virens*): Small-sized tree, scattered along the coast from Virginia to Texas.

102. LIVE OAK (*Quercus chrysolepis*). (maul oak, Valparaiso oak): Medium-sized tree; California.

OSAGE ORANGE.

103. OSAGE ORANGE (*Maclura aurantiaca*) (Bois d'Arc): Wood very heavy, exceedingly hard, strong, not tough, of moderately coarse texture, and very durable; sapwood yellow, heart brown on the end, yellow on longitudinal faces, soon turning grayish brown if exposed; it shrinks considerably in drying, but once dry it stands unusually well. Formerly much used for wheel stock in the dry regions of Texas; otherwise employed for posts, railway ties, etc. Seems too little appreciated; it is well suited for turned ware and especially for wood carving. A small-sized tree, of fairly rapid growth, scattered through the rich bottoms of Arkansas and Texas.

PERSIMMON.

104. PERSIMMON (*Diospyros virginiana*): Wood very heavy and hard, strong and tough; resembles hickory, but is of finer texture; the broad sapwood cream color, the heart black; used in turnery for shuttles, plane stocks, shoe lasts, etc. Small to medium sized tree, common and best developed in the Lower Ohio Valley, but occurs from New York to Texas and Missouri.

POPLAR AND COTTONWOOD (*See also* TULIP WOOD).—Wood light, very soft, not strong, of fine texture and whitish, grayish to yellowish color, usually with a satiny luster. The wood shrinks moderately (some crossgrained forms warp excessively), but checks little; is easily worked, but is not durable. Used as building and furniture lumber, in cooperage for sugar and flour barrels, for crates and boxes (especially cracker boxes), for wooden ware and paper pulp.

105. COTTONWOOD (*Populus monilifera*): Large sized tree; forms considerable forests along many of the Western streams, and furnishes most of the cottonwood of the market. Mississippi Valley and west; New England to the Rocky Mountains.

106. BALSAM (*Populus balsamifera*) (balm of Gilead): Medium to large sized tree; common all along the northern boundary of the United States.

107. BLACK COTTONWOOD (*Populus trichocarpa*): The largest deciduous tree of Washington; very common. Northern Rocky Mountains and Pacific region.

108. COTTONWOOD (*Populus fremontii* var. *wislizeni*): Medium to large sized tree, common. Texas to California.

109. POPLAR (*Populus grandidentata*): Medium-sized tree, chiefly used for pulp. Maine to Minnesota and southward along the Alleghanies.

110. ASPEN (*Populus tremuloides*): Small to medium sized tree, often forming extensive forests and covering burned areas. Maine to Washington and northward, south in the Western mountains to California and New Mexico.

SOUR GUM. (*See* GUM.)

RED GUM. (*See* GUM.)

SASSAFRAS.

111. SASSAFRAS (*Sassafras sassafras*): Wood light, soft, not strong, brittle, of coarse texture, durable; sapwood yellow, heart orange brown. Used in cooperage, for skiffs, fencing, etc. Medium-sized tree, largest in the Lower Mississippi Valley, from New England to Texas and from Michigan to Florida.

SWEET GUM. (*See* GUM.)

SYCAMORE.

112. SYCAMORE (*Platanus occidentalis*) (button wood, button-ball tree, water beech): Wood moderately heavy, quite hard, stiff, strong, tough, usually crossgrained, of coarse texture, and white to light brown color; the wood is

hard to split and work, shrinks moderately, warps and checks considerably, but stands well. It is used extensively for drawers, backs, bottoms, etc., in cabinet-work, for tobacco boxes, in cooperage, and also for finishing lumber, where it has too long been underrated. A large tree, of rapid growth, common and largest in the Ohio and Mississippi valleys, at home in nearly all parts of the Eastern United States. The California species—

113. *Platanus racemosa* resembles in its wood the Eastern form.

TULIP WOOD.

114. TULIP TREE (*Liriodendron tulipifera*) (yellow poplar, white wood): Wood quite variable in weight, usually light, soft, stiff but not strong, of fine texture, and yellowish color; the wood shrinks considerably, but seasons without much injury; works and stands remarkably well. Used for siding, for paneling and finishing lumber in house, car, and ship building, for sideboards and panels of wagons and carriages; also in the manufacture of furniture, implements and machinery, for pump logs, and almost every kind of common wooden ware, boxes, shelving, drawers, etc. An ideal wood for the carver and toy man. A large tree, does not form forests, but is quite common, especially in the Ohio Basin; occurs from New England to Missouri and southward to Florida.

115. CUCUMBER TREE (*Magnolia acuminata*): A medium-sized tree, most common in the Southern Alleghanies, but distributed from New York to Arkansas, southward to Alabama and northward to Illinois. Resembling, and probably confounded with, tulip wood in the markets.

TUPELO. (*See* GUM.)

WALNUT.

116. BLACK WALNUT (*Juglans nigra*): Wood heavy, hard, strong, of coarse texture; the narrow sapwood whitish, the heartwood chocolate brown. The wood shrinks moderately in drying, works and stands well, takes a good polish, is quite hand-some, and has been for a long time the favorite cabinet wood in this country. Walnut, formerly used even for fencing, has become too costly for ordinary uses, and is to-day employed largely as a veneer, for inside finish and cabinetwork; also in turnery, for gunstocks, etc. Black walnut is a large tree, with stout trunk, of rapid growth, and was formerly quite abundant throughout the Alleghany region, occurring from New England to Texas, and from Michigan to Florida.

WHITE WALNUT. (*See* BUTTERNUT.)

WHITE WOOD. (*See* TULIP, and also BASSWOOD.)

YELLOW POPLAR. (*See* TULIP.)

INDEX.

www.ingramcontent.com/pod-product-compliance
Lightning Source LLC
Chambersburg PA
CBHW020303090426
42735CB00009B/1194